Seismic Loads

About the International Code Council®

The International Code Council Inc. (ICC) is a nonprofit association that provides a wide range of building safety solutions including product evaluation, accreditation, certification, codification and training. It develops model codes and standards used worldwide to construct safe, sustainable, affordable and resilient structures. The mission of the Code Council is to provide the highest quality codes, standards, products and services for all concerned with the safety and performance of the built environment. ICC Evaluation Service (ICC-ES) is the industry leader in performing technical evaluations for code compliance fostering safe and sustainable design and construction.

S. K. Ghosh Associates LLC is a member of the ICC Family of Solutions with specialty in seismic and code consulting services providing technical support through publications, seminars, peer reviews, research projects, computer programs, and other support services. www.skghoshassociates.com; (847) 991-2700

ICC Washington DC Headquarters:
500 New Jersey Avenue, NW, 6th Floor, Washington, DC 20001

ICC Regional Offices:
Eastern Regional Office (BIR)
Central Regional Office (CH)
Western Regional Office (LA)
Distribution Center (Lenexa, KS)

888-ICC-SAFE (888-422-7233)
www.iccsafe.org

Seismic Loads
Time-Saving Methods Using the 2018 IBC and ASCE/SEI 7-16

David A. Fanella,
Ph.D., S.E., P.E., F.ACI, F.ASCE, F.SEI

New York Chicago San Francisco
Athens London Madrid
Mexico City Milan New Delhi
Singapore Sydney Toronto

Library of Congress Cataloging-in-Publication Data

Seismic Loads:
Time-Saving Methods Using the 2018 IBC and ASCE/SEI 7-16

1 2 3 4 5 6 7 8 9 CCD 25 24 23 22 21

ISBN 978-1-260-46739-0
MHID 1-260-46739-2

Sponsoring Editor
Ania Levinson

Editorial Supervisor
Donna M. Martone

Acquisitions Coordinator
Elizabeth Houde

Project Manager
Parag Mittal, KnowledgeWorks Global Ltd.

Copy Editor
KnowledgeWorks Global Ltd.

Proofreader
KnowledgeWorks Global Ltd.

Production Supervisor
Pamela A. Pelton

Composition
KnowledgeWorks Global Ltd.

Art Director, Cover
Jeff Weeks

About the Author

David. A. Fanella is Senior Director of Engineering at the Concrete Reinforcing Steel Institute where his main responsibility is creating educational material for structural engineers, including publications, design aids, and webinars. He has over 30 years of experience in a wide variety of low-, mid-, and high-rise buildings and other structures and has authored numerous books and technical papers through the years, including two editions of *Reinforced Concrete Structures, Analysis and Design*. David is a licensed Structural Engineer and Professional Engineer in Illinois, and is a Fellow of the American Concrete Institute, the American Society of Civil Engineers, and the Structural Engineers Institute. He is active in many professional organizations, including membership on ASCE/SEI 7 and ACI Committees. He is also past President and past Board Member of the Structural Engineers Association of Illinois.

Contents

Preface

This publication provides structural engineers, educators, students, and other design professionals a concise, visual guide to the determination of structural loads due to seismic loads. The intent is to present the provisions in the 2018 *International Building Code* and ASCE/SEI 7-16 *Minimum Design Loads and Associated Criteria for Buildings and Other Structures* in a manner that is easy to understand and apply. This is achieved by utilizing step-by-step methods including numerous figures, tables, flowcharts, and design aids.

Examples in both inch-pound and metric (S.I.) units illustrate the proper application of the code provisions and follow the step-by-step methods outlined throughout the publication. Section, figure, table, and equation numbers from the code and this publication are given in the right-hand margin of the examples for easy reference.

In short, seismic loads can be determined simpler and faster using the procedures in this publication.

For further online information on this topic, please go to https://www.mhprofessional .com/SeismicLoads

David A. Fanella

Seismic Loads

CHAPTER 1

Introduction

1.1 Overview

The purpose of this publication is to assist in the proper determination of seismic loads in accordance with the 2018 edition of the *International Building Code*© (IBC©) [Ref. 1; see Chap. 6 of this publication for a list of references] and the 2016 edition of ASCE/SEI 7 *Minimum Design Loads and Associated Criteria for Buildings and Other Structures* (Ref. 2), including Supplement 1, which became effective in December of 2018 (https://ascelibrary .org/doi/10.1061/9780784414248.sup1). The main goal is to streamline the load determination process by providing straightforward, step-by-step procedures enhanced by numerous design aids, figures, and flowcharts, which provide a roadmap through the numerous code requirements.

Design professionals will appreciate the simplicity and thoroughness of the content and will find the "how to" methods of load determination useful in everyday practice. Worked-out examples illustrate the proper application of the code requirements and follow the step-by-step procedures noted above; these examples are a valuable resource for individuals studying for licensing exams, undergraduate and graduate students, and others involved in structural engineering.

Readers interested in the background, history, and design philosophy of the code requirements for seismic loads can find detailed information and references in the commentary of Ref. 2 and in Ref. 3.

1.2 Scope

Throughout this publication, section numbers from the IBC are referenced as illustrated by the following: Section 1613 of the IBC is denoted as IBC 1613. Similarly, Section 11.1 of ASCE/SEI 7-16 is referenced as ASCE/SEI 11.1.

Chapter 2 contains the seismic design criteria for the design and construction of buildings and other structures subjected to earthquake ground motions. Information is provided on how to determine seismic ground motion values, risk category and seismic importance factor, seismic design category, design requirements for seismic design category A, geological hazards and geotechnical investigations, and vertical ground motions.

Seismic design requirements for building structures are covered in Chap. 3. Methods are presented on how to determine design seismic forces for seismic force resisting systems, diaphragms, chords, collectors, and structural walls and their anchorages. Also covered are structural irregularities, redundancy, seismic load effects and combinations, direction of loading, modeling criteria, and drift and deformation.

3

Chapter 4 contains the seismic design requirements for nonstructural components, including architectural components, mechanical and electrical components, and anchorage. Seismic design requirements for nonbuilding structures are given in Chap. 5. Included are methods on how to determine design seismic forces for nonbuilding structures similar to buildings and nonbuilding structures that are not similar to buildings. Chapter 6 contains the references cited in this publication.

Both inch-pound and S.I. units are used throughout this publication, including in the equations, figures, tables, flow charts, and examples. In the examples, calculations are performed independently using both sets of units; in other words, the calculations are not performed in one set of units and then converted to the other. Thus, in some cases, the numerical results in inch-pound units do not "exactly" convert to the corresponding numerical results in S.I. units or vice versa.

1.3   Notation

A_b = area of base of a structure, ft^2 (m^2)

A_i = web area of a shear wall i, ft^2 (m^2)

A_x = torsional amplification factor

a_p = amplification factor related to the response of a system or component as affected by the type of seismic attachment

C_d = deflection amplification factor

C_{dX} = deflection amplification factor in the x-direction

C_{dY} = deflection amplification factor in the y-direction

C_{pi} = diaphragm design acceleration coefficient at 80 percent of the structural height above the base

C_{pn} = diaphragm design acceleration coefficient at the structural height above the base

C_{px} = diaphragm design acceleration coefficient at level x

C_{p0} = diaphragm design acceleration coefficient at the structure base

C_R = site-specific risk coefficient at any period

C_{RS} = mapped value of the risk coefficient at short periods

C_{R1} = mapped value of the risk coefficient at a period of 1 s

C_s = seismic response coefficient

C_{s2} = higher mode seismic response coefficient

C_t = building period coefficient

C_{vx} = vertical distribution factor

C_w = parameter used in calculating the approximate fundamental period of structures with masonry or concrete shear walls

D = effect of dead load

D_i = length of shear wall i, ft (m)

D_{pl} = seismic relative displacement

D_s = total depth of stratum, ft (m)

d_c = total thickness of cohesive soil layers in the top 100 ft (30 m), ft (m)

d_i = thickness of any soil or rock layer i between 0 and 100 ft (30 m), ft (m)

d_s = total thickness of cohesionless soil layers in the top 100 ft (30 m), ft (m)

E = effect of horizontal and vertical earthquake-induced forces

E_{cl} = capacity-limited horizontal seismic load effect, equal to the maximum force that can develop in the element as determined by a rational, plastic mechanism analysis

E_h = horizontal seismic load effect

E_{mh} = horizontal seismic load effect including overstrength

E_v = vertical seismic load effect

F_a = short-period site coefficient at 0.2-s period

F_i = portion of the seismic base shear, V, induced at level i, lb (N)

F_n = portion of the seismic base shear, V, induced at level n, lb (N)

F_x = portion of the seismic base shear, V, induced at level x, lb (N)

F_{PGA} = site coefficient for peak ground acceleration (PGA)

F_p = seismic force acting on a component of a structure, lb (N)

F_{px} = diaphragm seismic force at level x, lb (N)

F_v = long-period site coefficient at 1.0-s period

F_x = portion of the seismic base shear, V, induced at level x, lb (N)

G = average shear modulus for the soils beneath the foundation at large strain levels, lb/ft^2 (N/m^2)

g = acceleration due to gravity

H = thickness of soil, ft (m)

h = average roof height of a structure with respect to the base, ft (m)

h_i = height above the base to level i, ft (m)

h_n = vertical distance from the base to the highest level of the seismic force-resisting system of the structure; for pitched or sloped roofs, the structure height is from the base to the average height of the roof, ft (m)

h_{sx} = story height below level x, ft (m)

h_x = height above the base to level x, ft (m)

I_e = seismic importance factor

I_p = component importance factor

k = exponent related to the period of a structure

k_a = coefficient defined in ASCE/SEI 12.11.2.1 for wall anchorage forces and in ASCE/SEI 12.14.7.5 for anchorage of structural walls

L = overall length of a building at the base in the direction of analysis, ft (m)

L_f = span of a flexible diaphragm providing lateral support for the wall, which is measured between vertical elements that provide lateral support to the diaphragm in the direction of analysis; use zero for rigid diaphragms, ft (m)

M_t = torsional moment resulting from eccentricity between the locations of center of mass and center of rigidity, ft-lb (N-m)

M_{ta} = accidental torsional moment, ft-lb (N-m)

MCE = maximum considered earthquake ground motion, %g

MCE_G = maximum considered earthquake geometric mean peak ground acceleration, %g

MCE_R = risk-targeted maximum considered earthquake ground motion response acceleration, %g

N = standard penetration resistance, blows/ft (blows/m)

   = number of stories above the base

$\overline{N}$ = average field standard penetration resistance for the top 100 ft (30 m), blows/ft (blows/m)

$\overline{N}_{ch}$ = average standard penetration resistance for cohesionless soil layers for the top 100 ft (30 m), blows/ft (blows/m)

N_i = standard penetration resistance of any soil or rock layer i between 0 and 100 ft (30 m), blows/ft (blows/m)

P_x = total unfactored vertical design load at and above level x, lb (N)

PGA = mapped MCE_G peak ground acceleration, %g

PGA_M = MCE_G peak ground acceleration adjusted for site class effects, %g

PI = plasticity index

Q_E = effect of horizontal seismic (earthquake-induced) forces

R = response modification coefficient

R_p = component response modification coefficient

R_s = diaphragm design force reduction factor

R_X = response modification coefficient in the x-direction

R_Y = response modification coefficient in the y-direction

S_{aM} = site-specific MCE_R spectral response acceleration parameter at any period, %g

S_{D1} = design, 5% damped, spectral response acceleration parameter at a period of 1 s, %g

S_{DS} = design, 5% damped, spectral response acceleration parameter at short periods, %g

S_{M1} = the MCE_R, 5% damped, spectral response acceleration parameter at a period of 1 s adjusted for site class effects, %g

S_{MS} = the MCE_R, 5% damped, spectral response acceleration parameter at short periods adjusted for site class effects, %g

S_S = mapped MCE_R, 5% damped, spectral response acceleration parameter at short periods, %g

S_1 = mapped MCE_R, 5% damped, spectral response acceleration parameter at a period of 1 s, %g

s_u = undrained shear strength, lb/ft² (N/m²)

$\bar{s}_u$ = average undrained shear strength in the top 100 ft (30 m), lb/ft² (N/m²)

s_{ui} = undrained shear strength of any cohesive soil layer i between 0 and 100 ft (30 m), lb/ft² (N/m²)

T = fundamental period of a building, s

T_a = approximate fundamental period of a building, s

T_L = long-period transition period, s

T_p = fundamental period of the component and its attachment, s

$T_S = S_{D1}/S_{DS}$

$T_0 = 0.2 S_{D1}/S_{DS}$

V = total design lateral force or shear at the base, lb (N)

V_x = seismic design story shear in any story, lb (N)

v_s = shear wave velocity at small shear strains, ft/s (m/s)

$\bar{v}_s$ = average shear wave velocity at small shear strains in the top 100 ft (30 m), ft/s (m/s)

v_{si} = shear wave velocity of any soil or rock layer i between 0 and 100 ft (30 m), ft/s (m/s)

W = effective seismic weight of a building, lb (N)

W_c = gravity load of a component of a building, lb (N)

W_p = component operating weight, lb (N)

w = moisture content (in percent)

w_i = portion of W located at or assigned to level i, lb (N)

w_n = portion of W located at or assigned to level n, lb (N)

w_{px} = weight tributary to a diaphragm at level x, lb (N)

w_x = portion of W located at or assigned to level x, lb (N)

x = level under consideration

= number of shear walls in the building effective in resisting lateral forces in the direction of analysis

z = height in structure at point of attachment of a component with respect to the base, ft (m)

z_s = mode shape factor

β = ratio of shear demand to shear capacity for the story between levels x and $x - 1$

Γ_{m1} = first modal contribution factor

Γ_{m1} = higher modal contribution factor

γ = average unit weight of soil, lb/ft³ (N/m³)

Δ = design story drift, in. (mm)

Δ_{ADVE} = average drift of adjoining vertical elements of the seismic force–resisting system over the story below the diaphragm under consideration, subjected to a tributary lateral load equivalent to that use in the computation of δ_{MDD}, in. (mm)

Δ_a = allowable story drift, in. (mm)

δ_M = maximum inelastic response displacement, considering torsion, in. (mm)

δ_{MDD} = computed maximum in-plane deflection of a diaphragm under lateral load, in. (mm)

δ_{MT} = total separation distance between adjacent structures on the same property, in. (mm)

δ_{avg} = average of the displacements at the extreme points of the same structure at level x, in. (mm)

δ_{max} = maximum displacement at level x, considering torsion, in. (mm)

δ_x = deflection of level x at the center of mass at and above level x, in. (mm)

δ_{xc} = deflection of level x at the center of mass at and above level x determined by an elastic analysis, in. (mm)

θ = stability coefficient for P-delta effects

θ_{max} = maximum permitted stability coefficient

ρ = redundancy factor

Ω_o = overstrength factor

Ω_v = diaphragm shear overstrength factor

CHAPTER 2

Seismic Design Criteria

2.1 Overview

According to IBC 1613.1, every structure and portion thereof, including nonstructural components permanently attached to structures and their supports and attachments, must be designed and constructed to resist the effects of earthquake motions in accordance with ASCE/SEI 7-16 Chapters 11, 12, 13, 15, 17, and 18, as applicable. The following structures are exempt from the seismic requirements (IBC 1613.1 and ASCE/SEI 11.1.2):

1. Detached one- and two-family dwellings assigned to Seismic Design Category (SDC) A, B, or C, or located where the mapped short-period spectral response acceleration, S_S, is less than 0.4.

2. Detached one- and two-family wood-frame dwellings not included in exemption 1 with not more than 2 stories above grade and satisfying the limitations of and constructed in accordance with Ref. 4.

3. Agricultural storage structures intended only for incidental human occupancy.

4. Structures requiring special consideration of their response characteristics and environment not covered in ASCE/SEI Chapter 15 and for which other regulations provide seismic criteria, such as vehicular bridges, electrical transmission towers, hydraulic structures, buried utility lines and their appurtenances, and nuclear reactors.

5. Piers and wharves not accessible to the general public.

This chapter contains the criteria for the design and construction of buildings and other structures subjected to earthquake ground motions. The following are covered in this chapter:

- Seismic ground motion values (Sec. 2.2)
- Importance factor and risk category (Sec. 2.3)
- Seismic design category (Sec. 2.4)
- Design requirements for SDC A (Sec. 2.5)
- Geological hazards and geotechnical investigation (Sec. 2.6)
- Vertical ground motions for seismic design (Sec. 2.7)

2.2 Seismic Ground Motion Values

2.2.1 Near-Fault Sites

Structures located in close proximity to the zone of fault rupture can be subjected to very damaging ground motions. More restrictive design criteria must be satisfied in ASCE/SEI 7 where such ground motion can occur.

Sites satisfying either of the following two conditions are classified as near-fault sites (see ASCE/SEI 11.4.1 and Fig. 2.1):

- 9.5 miles (15 km) of the surface projection of a known active fault capable of producing M_w7 or larger events.

- 6.25 miles (10 km) of the surface projection of a known active fault capable of producing M_w6 or larger events.

The moment magnitude scale denoted by M_w is a measure of an earthquake's magnitude (that is, size or strength) based on its seismic moment, which is a measure of the work done by the faulting of an earthquake.

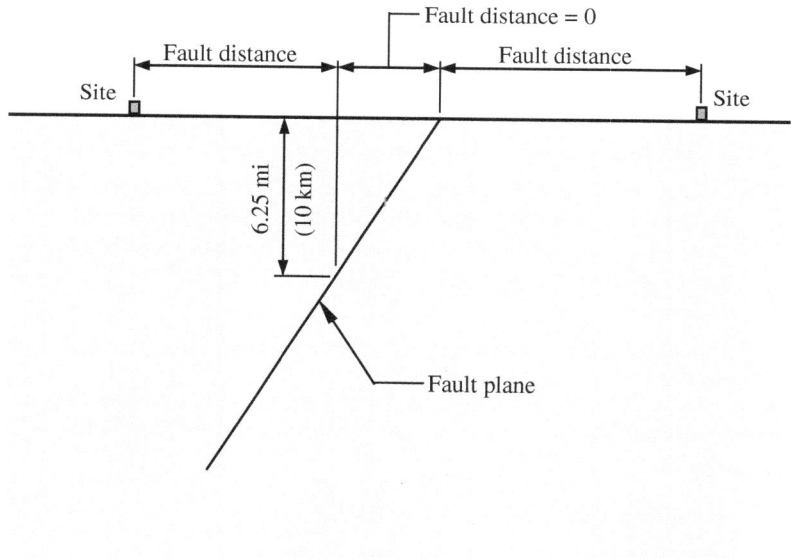

Near-fault site:

$$\text{Fault distance} \leq \begin{cases} 9.5 \text{ mi (15 km) from an } M_w7 \text{ or larger active fault} \\ 6.25 \text{ mi (10 km) from an } M_w6 \text{ or larger active fault} \end{cases}$$

Figure 2.1 Near-fault criteria of ASCE/SEI 11.4.1.

Two exceptions are also given in ASCE/SEI 11.4.1:

1. Faults with estimated slip rates along the fault less than 0.04 in. (1 mm) per year need not be considered.

2. Surface projections need not include portions of the fault at depths of 6.25 miles (10 km) or greater (see Fig. 2.1).

2.2.2 Mapped Acceleration Parameters

Contour maps of S_S and S_1, which are the risk-targeted maximum considered earthquake (MCE_R) ground motion parameters (response accelerations) at periods of 0.2 s and 1 s, respectively, for sites with an effective average small-strain shear wave velocity, $\bar{v}_s$, of 2,500 ft/s (760 m/s) and 5-percent damping, are given in the IBC and ASCE/SEI 7 (see Table 2.1). The MCE_R earthquake corresponds to probabilistic ground motion (that is, ground motion with a mean recurrence interval of 2,475 years) except at locations where deterministic ground motion governs (see Ref. 3). This reference earthquake shaking level provides a small probability (approximately 10 percent or less) that structures with ordinary occupancies will collapse when subjected to such shaking.

The short-period acceleration, S_S, has been determined at a period of 0.2 s because it was concluded that 0.2 s was reasonably representative of the shortest effective period of buildings and structures designed using these requirements. The 1-s acceleration, S_1, is used because spectral response accelerations at periods other than 1 s typically can be derived from the acceleration at 1 s.

In lieu of the contour maps, which can be very imprecise due to their large scale, S_S and S_1 may be obtained from Refs. 5 through 7 by entering either an address or latitude and longitude coordinates of a site.

Where $S_S \leq 0.15$ and $S_1 \leq 0.04$, a building or structure is permitted to be assigned to SDC A and must only satisfy the requirements of ASCE/SEI 11.7 (ASCE/SEI 11.4.2).

| | S_S | | S_1 | |
| | Figure No. | | Figure No. | |
Location	IBC	ASCE/SEI 7	IBC	ASCE/SEI 7
Conterminous U.S.	1613.2.1(1)	22-1	1613.2.1(2)	22-2
Hawaii	1613.2.1(3)	22-5	1613.2.1(3)	22-5
Alaska	1613.2.1(4)	22-3	1613.2.1(5)	22-4
Puerto Rico U.S. Virgin Islands	1613.2.1(6)	22-6	1613.2.1(6)	22-6
Guam Northern Mariana Islands	1613.2.1(7)	22-7	1613.2.1(7)	22-7
American Samoa	1613.2.1(8)	22-8	1613.2.1(8)	22-8

TABLE 2.1 Summary of Mapped Acceleration Parameters S_S and S_1 in the 2018 IBC and ASCE/SEI 7-16

2.2.3   Site Class

Six site classes are defined in ASCE/SEI 20.3 (see Table 2.2). A site is classified based on the following soil properties measured over the top 100 ft (30 m) of the site (ASCE/SEI 20.1):

1. Effective average small-strain shear wave velocity, $\bar{v}_s$ (ASCE/SEI 20.4.1)
2. Average field standard penetration resistance, $\bar{N}$, or average standard penetration resistance for cohesionless soil layers, $\bar{N}_{ch}$ (ASCE/SEI 20.4.2)
3. Average undrained shear strength, $\bar{s}_u$ (ASCE/SEI 20.4.3)

Definitions of the site class parameters are given in Table 2.3 (ASCE/SEI 20.4).

Site class parameters are usually determined by a geotechnical investigation using soil samples collected at various locations in the site. Where site-specific data are not available to a depth of 100 ft (30 m) or where it may not be possible to drill the entire 100-ft (30-m) depth to acquire soil properties (for example, a layer of hard rock is encountered 25 ft (7.6 m) below the ground surface), soil properties are permitted to be estimated by the registered design professional preparing the soil investigation report based on known geological conditions (ASCE/SEI 20.1). Where soil properties are not

Site Class	$\bar{v}_s$, ft/s (m/s)	$\bar{N}$ or $\bar{N}_{ch}$*	$\bar{s}_u$, lb/ft² (kN/m²)
A—Hard rock	>5,000 (>1,524)	Not applicable	Not applicable
B—Rock	2,500 to 5,000 (762 to 1,524)	Not applicable	Not applicable
C—Very dense soil and soft rock	1,200 to 2,500 (366 to 762)	>50	>2,000 (>96)
D—Stiff soil	600 to 1,200 (183 to 366)	15 to 50	1,000 to 2,000 (48 to 96)
E—Soft clay soil	<600 (<183)	<15	<1,000 (<48)
	Any profile with more than 10 ft (3.1 m) of soil with the following characteristics: • Plasticity index $PI > 20$ • Moisture content $w \geq 40\%$ • Undrained shear strength $\bar{s}_u < 500$ lb/ft² (<24 kN/m²)		
F—Soils requiring site response analysis in accordance with ASCE/SEI 21.1	See ASCE/SEI 20.3.1		

*Number of blows per 12 in. (305 mm)

TABLE 2.2   Site Classification in Accordance with ASCE/SEI 20.3

Parameter	Definition	ASCE/SEI Equation No.
Effective average small-strain shear wave velocity, $\bar{v}_s$	$\bar{v}_s = \sum_{i=1}^{n} d_i \Big/ \sum_{i=1}^{n}(d_i/v_{si})$	20.4-1
Average field standard penetration resistance, $\bar{N}$	$\bar{N} = \sum_{i=1}^{n} d_i \Big/ \sum_{i=1}^{n}(d_i/N_i)$	20.4-2
Average standard penetration resistance for cohesionless soil layers, $\bar{N}_{ch}$	$\bar{N}_{ch} = d_s \Big/ \sum_{i=1}^{m}(d_i/N_i)$	20.4-3
Average undrained shear strength, $\bar{s}_u$	$\bar{s}_u = d_c \Big/ \sum_{i=1}^{k}(d_i/s_{ui})$	20.4-4

(1) d_i = thickness of any layer between 0 and 100 ft (30 m); $\sum_{i=1}^{n} d_i = 100$ ft (30 m)
(2) v_{si} = shear wave velocity in ft/s (m/s)
(3) In ASCE/SEI Equation (20.4-2), N_i and d_i are for cohesionless soil, cohesive soil, and rock layers where N_i = standard penetration resistance not to exceed 100 blows/ft (305 blows/m)
(4) In ASCE/SEI Equation (20.4-3), N_i and d_i are for cohesionless soil layers only where N_i = standard penetration resistance not to exceed 100 blows/ft (305 blows/m)
(5) In ASCE/SEI Equation (20.4-3), d_s = total thickness of cohesionless soil layers in the top 100 ft (30 m) where $\sum_{i=1}^{m} d_i = d_s$
(6) In ASCE/SEI Equation (20.4-4), d_c = total thickness of cohesive soil layers in the top 100 ft (30 m); PI = plasticity index; w = moisture content in percent; s_{ui} = undrained shear strength in lb/ft² (kN/m²) not to exceed 5,000 lb/ft² (240 kN/m²); and $\sum_{i=1}^{k} d_i = d_c$

TABLE 2.3 Definitions of Site Class Parameters

known in sufficient detail to determine the site class, Site Class D, subject to the requirements in ASCE/SEI 11.4.4, must be used, unless the authority having jurisdiction requires or has reason to believe the site would be more properly classified as Site Class E or F (ASCE/SEI 11.4.3 and 20.1).

The flowchart in Fig. 2.2 can be used to determine the site class of a site in accordance with ASCE/SEI Chapter 20.

2.2.4 Site Coefficients and *MCE*ᴿ Spectral Response Acceleration Parameters

The MCE_R spectral response acceleration parameters for short periods, S_{MS}, and at 1 s periods, S_{M1}, adjusted for site class effects, are determined by ASCE/SEI Equations (11.4-1) and (11.4-2), respectively:

$$S_{MS} = F_a S_S \qquad (2.1)$$

$$S_{M1} = F_v S_1 \qquad (2.2)$$

Site coefficients F_a and F_v are given in ASCE/SEI Tables 11.4-1 and 11.4-2, respectively (see Supplement 1 for the updated versions of these tables). Tables 2.4 and 2.5 are similar to ASCE/SEI Tables 11.4-1 and 11.4-2; extra rows have been added to account for the requirements in ASCE/SEI 11.4.3 and 11.4.4 (see Table 2.6).

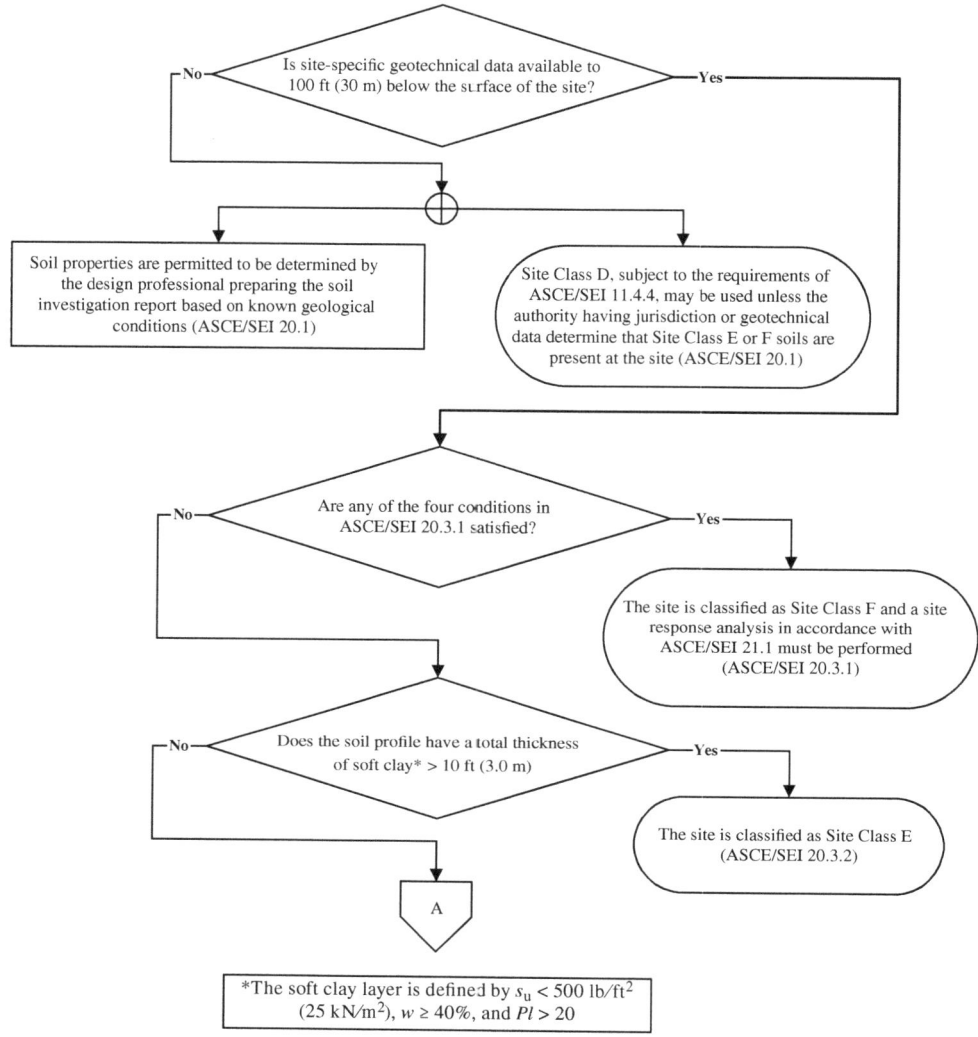

FIGURE 2.2   Site classification procedure in accordance with ASCE/SEI Chapter 20.

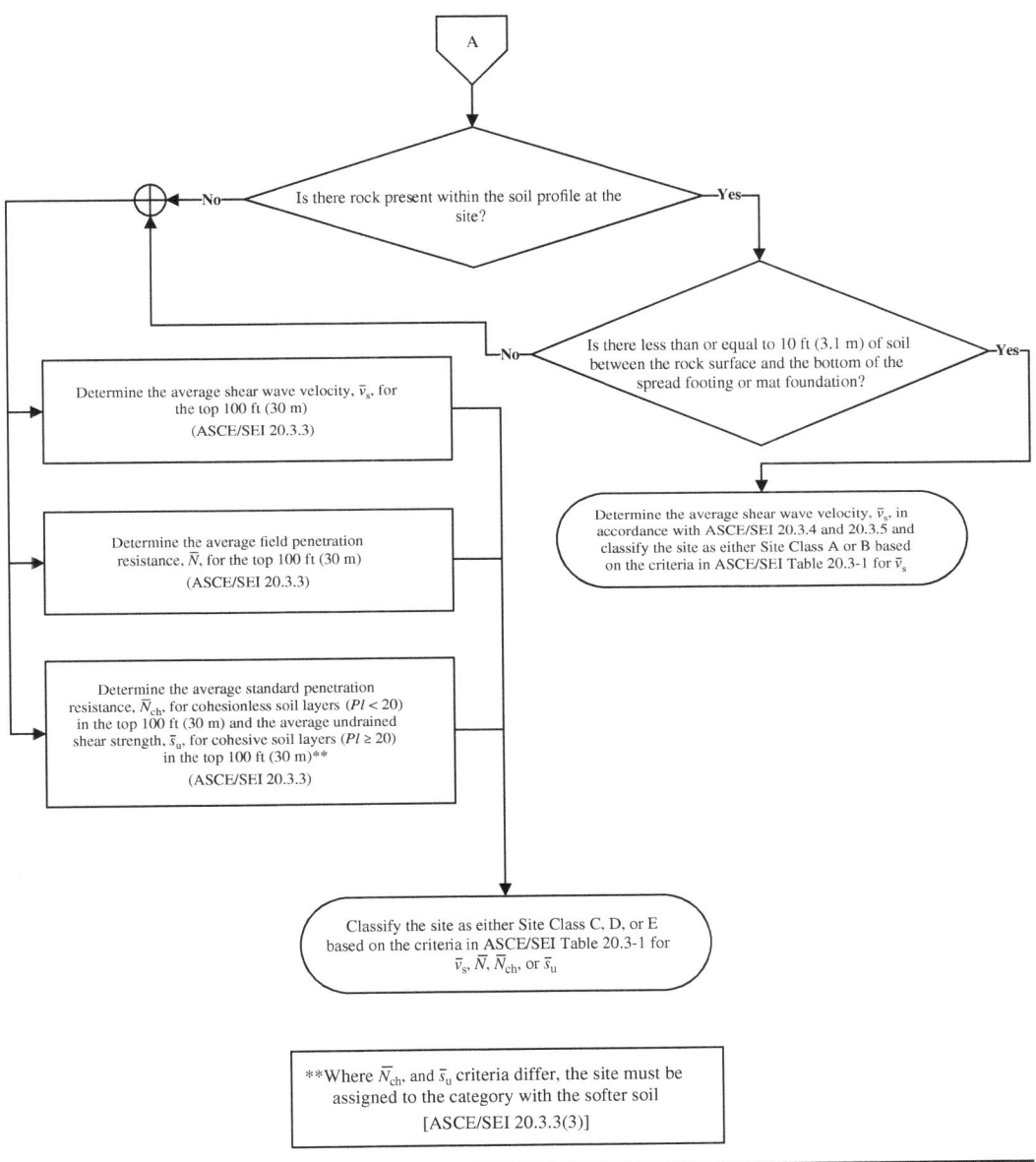

FIGURE 2.2 (Continued)

Site Class	MCE_R Spectral Response Acceleration Parameter at Short Period					
	$S_s \leq 0.25$	$S_s = 0.50$	$S_s = 0.75$	$S_s = 1.00$	$S_s = 1.25$	$S_s \geq 1.50$
A	0.8	0.8	0.8	0.8	0.8	0.8
B—Rock	0.9	0.9	0.9	0.9	0.9	0.9
B—Estimated	1.0	1.0	1.0	1.0	1.0	1.0
C	1.3	1.3	1.2	1.2	1.2	1.2
D—Stiff Soil	1.6	1.4	1.2	1.1	1.0	1.0
D—Default	1.6	1.4	1.2	1.2	1.2	1.2
E	2.4	1.7	1.3	(a)	(a)	(a)
F	(a)	(a)	(a)	(a)	(a)	(a)

(a) See ASCE/SEI 11.4.8.

TABLE 2.4   Short-Period Site Coefficient, F_a

Site Class	MCE_R Spectral Response Acceleration Parameter at 1-s Period					
	$S_1 \leq 0.10$	$S_1 = 0.20$	$S_1 = 0.30$	$S_1 = 0.40$	$S_1 = 0.50$	$S_1 \geq 0.60$
A	0.8	0.8	0.8	0.8	0.8	0.8
B—Rock	0.8	0.8	0.8	0.8	0.8	0.8
B—Estimated	1.0	1.0	1.0	1.0	1.0	1.0
C	1.5	1.5	1.5	1.5	1.5	1.4
D—Stiff Soil	2.4	2.2[b]	2.0[b]	1.9[b]	1.8[b]	1.7[b]
D—Default	2.4	2.2[b]	2.0[b]	1.9[b]	1.8[b]	1.7[b]
E	4.2	3.3[b]	2.8[b]	2.4[b]	2.2[b]	2.0[b]
F	(a)	(a)	(a)	(a)	(a)	(a)

(a) See ASCE/SEI 11.4.8.
(b) See requirements for site-specific ground motions in ASCE/SEI 11.4.8. These values of F_v are to be used only for the calculation of T_s.

TABLE 2.5   Long-Period Site Coefficient, F_v

2.2.5   Design Spectral Acceleration Parameters

Design earthquake spectral response acceleration parameters for short periods, S_{DS}, and at 1 s periods, S_{D1}, are determined by ASCE/SEI Equations (11.4-3) and (11.4-4), respectively:

$$S_{DS} = 2S_{MS}/3 \tag{2.3}$$

$$S_{D1} = 2S_{M1}/3 \tag{2.4}$$

Designation	Definition	F_a and F_v
Site Class B—Rock	Site Class B has been established based on site investigations in accordance with ASCE/SEI Chapter 20 and site-specific shear wave velocity measurements have been made	F_a and F_v must be taken as 0.9 and 0.8, respectively
Class B—Estimated	Site Class B has been established based on site investigations in accordance with ASCE/SEI Chapter 20 and site-specific shear wave velocity measurements have not been made	F_a and F_v must be taken as 1.0 (ASCE/SEI 11.4.3)
Site Class D—Stiff Soil	Site Class D has been established based on site investigations in accordance with ASCE/SEI Chapter 20 and site-specific shear wave velocity measurements have been made	F_a and F_v in Tables 2.4 and 2.5 are to be used, respectively, subject to the requirements in ASCE/SEI 11.4.8
Site Class D—Default	Soil properties are not known in sufficient detail to determine the site class (ASCE/SEI 11.4.3)	F_a must be greater than or equal to 1.2 (ASCE/SEI 11.4.4)

TABLE 2.6   Site Coefficient Requirements of ASCE/SEI 11.4.3 and 11.4.4

Structural design in accordance with the IBC and ASCE/SEI 7 is performed for earthquake demands that are equal to two-thirds of the MCE_R ground motion.

The flowchart in Fig. 2.3 can be used to determine S_{DS} and S_{D1} for a site in accordance with ASCE/SEI 11.4.

2.2.6   Design Response Spectrum

The generalized form of the design response spectrum based on the design spectral response acceleration parameters discussed previously is illustrated in Fig. 2.4 (see ASCE/SEI Figure 11.4-1).

Mapped values of the long-period transition period, T_L, are given in ASCE/SEI Figures 22-14 through 22-17 (see Table 2.7). Alternatively, T_L can be obtained from Refs. 5 through 7.

The spectrum in Fig. 2.4 is to be used wherever required in ASCE/SEI 7 and should not be used in cases where the site-specific ground motion procedures in ASCE/SEI 11.4.8 must be used (ASCE/SEI 11.4.6). Requirements where a site response analysis is performed or required are given in ASCE/SEI Chapter 21 (see Sec. 2.2.8 of this publication).

2.2.7   Risk-Targeted MCE_R Response Spectrum

Where a risk-targeted MCE_R response spectrum is required, it is obtained by multiplying the spectral response accelerations determined by the design response spectrum in Sec. 2.2.6 of this publication by 1.5.

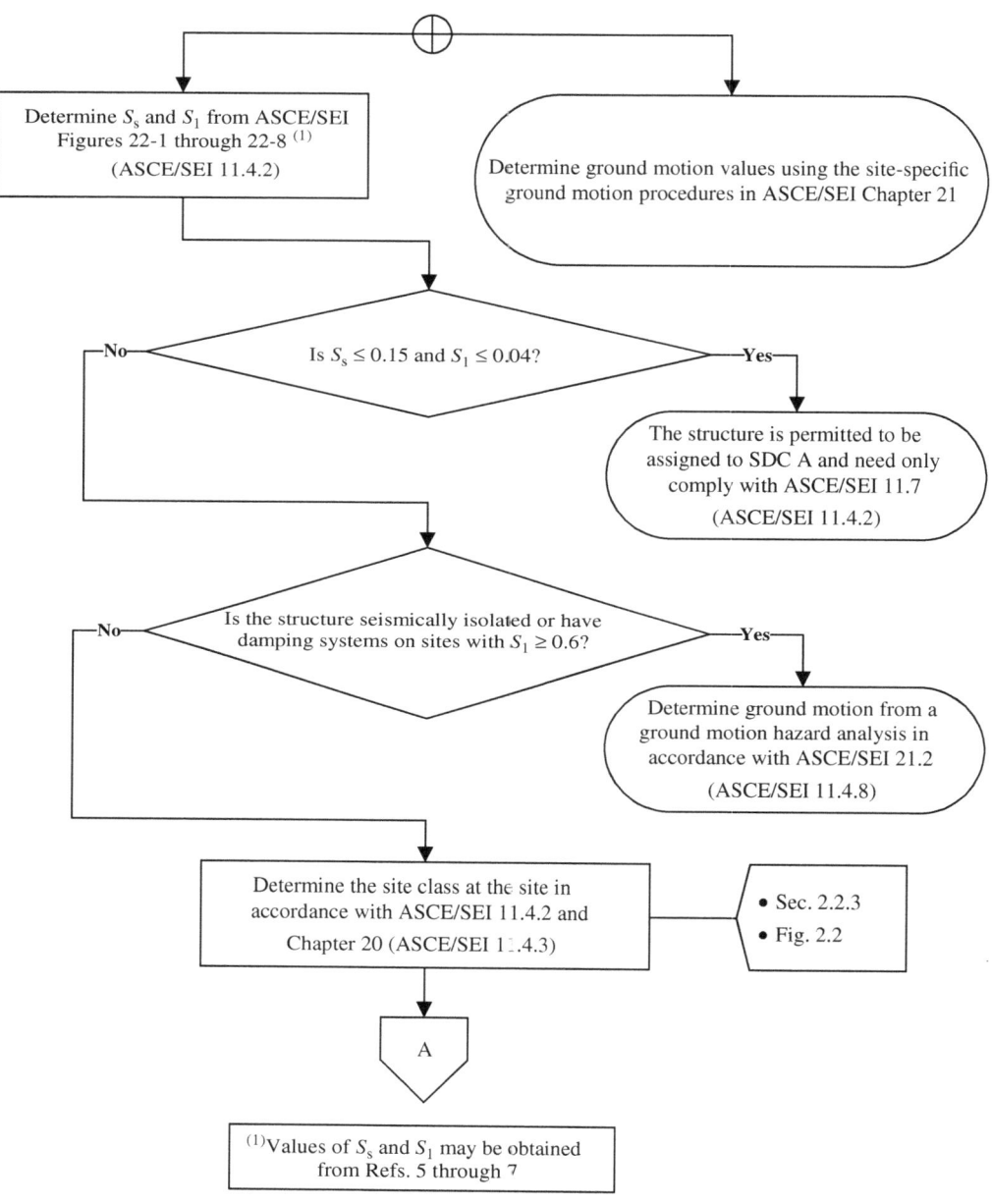

Determine S_s and S_1 from ASCE/SEI Figures 22-1 through 22-8 [1] (ASCE/SEI 11.4.2)

Determine ground motion values using the site-specific ground motion procedures in ASCE/SEI Chapter 21

Is $S_s \leq 0.15$ and $S_1 \leq 0.04$?

—No   Yes—

The structure is permitted to be assigned to SDC A and need only comply with ASCE/SEI 11.7 (ASCE/SEI 11.4.2)

Is the structure seismically isolated or have damping systems on sites with $S_1 \geq 0.6$?

—No   Yes—

Determine ground motion from a ground motion hazard analysis in accordance with ASCE/SEI 21.2 (ASCE/SEI 11.4.8)

Determine the site class at the site in accordance with ASCE/SEI 11.4.2 and Chapter 20 (ASCE/SEI 11.4.3)

- Sec. 2.2.3
- Fig. 2.2

A

[1]Values of S_s and S_1 may be obtained from Refs. 5 through 7

FIGURE 2.3   Seismic ground motion parameters.

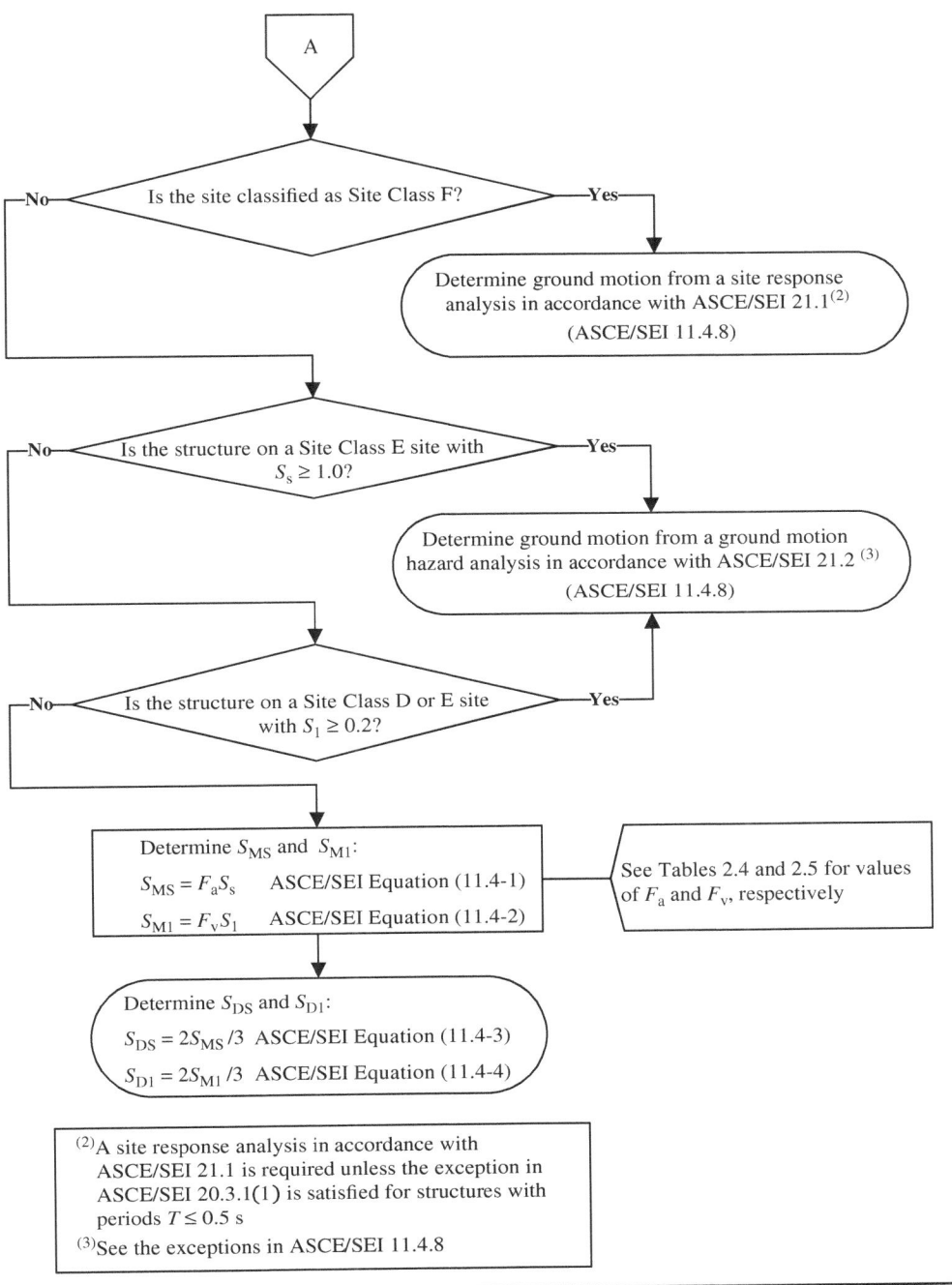

FIGURE 2.3 *(Continued)*

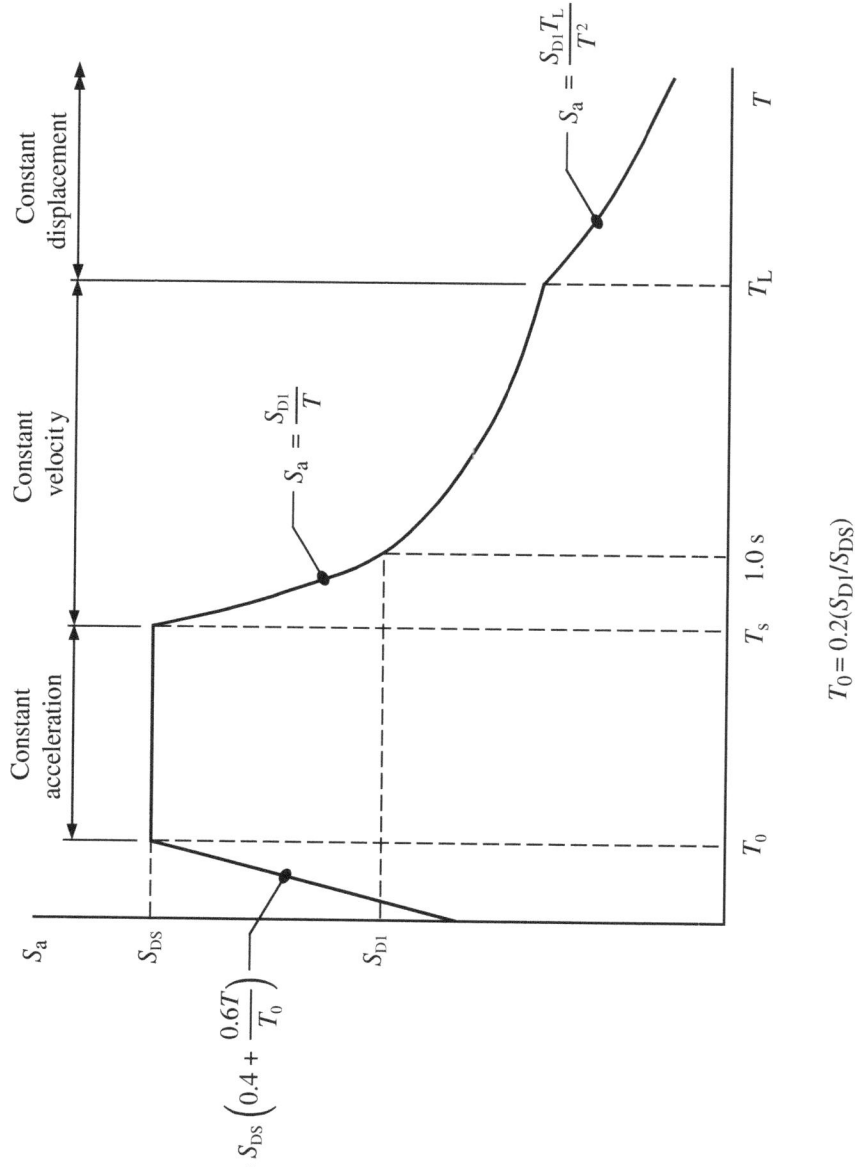

$$T_0 = 0.2(S_{D1}/S_{DS})$$

$$T_s = S_{D1}/S_{DS}$$

FIGURE 2.4 Design response spectrum in accordance with ASCE/SEI 11.4.6.

22

Location	ASCE/SEI Figure No.
Conterminous U.S.	22-14
Alaska	22-15
Hawaii	22-16
Puerto Rico U.S. Virgin Islands	22-17

TABLE 2.7 Summary of Mapped Long-Period Transition Period, T_L

2.2.8 Site-Specific Ground Motion Procedures

Overview

The site-specific ground motion procedures in ASCE/SEI Chapter 21 may be used to determine ground motions for any structure and are required for certain structures and certain site soil conditions. The objective of a site-specific ground motion analysis is to determine seismic ground motion values for local seismic and site conditions with higher confidence than is possible using the general procedure in ASCE/SEI 11.4 (ASCE/SEI C11.4.8).

A summary of the procedures that must be used to determine site-specific ground motions is given in Table 2.8 along with exceptions to the requirements.

The following requirements must be satisfied where the analysis methods of either ASCE/SEI 21.1 or 21.2 are used (ASCE/SEI 11.4.8):

- The design response spectrum must be determined in accordance with ASCE/SEI 21.3.
- The design acceleration parameters must be determined in accordance with ASCE/SEI 21.4.
- The MCE_G peak ground acceleration parameter must be determined in accordance with ASCE/SEI 21.5, where required.

Site Response Analysis

In a site response analysis, a model is developed using criteria established by a site-specific geotechnical investigation, which should include the following (ASCE/SEI C21.1.2): (1) borings with sampling; (2) standard penetration tests, cone penetrometer tests, and/or subsurface investigative techniques; and (3) laboratory testing to establish soil types, properties, and layering.

Base ground motion time histories are input to the soil profile model at the site as outcropping motions, and surface ground motion time histories are calculated. The calculated MCE_R ground motion response spectrum must be greater than or equal to the MCE_R response spectrum of the base motion multiplied by the average surface-to-base response spectral ratios (calculated period by period) obtained from the site response analyses (ASCE/SEI 21.1.3).

Condition	Required Analysis Method	Exceptions
Structures on Site Class F sites	Site response analysis in accordance with ASCE/SEI 21.1	• For structures with $T \leq 0.5$ s and located on sites with liquefiable soils, ground motions are permitted to be determined in accordance with ASCE/SEI 20.3 using F_a and F_v from ASCE/SEI Tables 11.4-1 and 11.4-2, respectively. • For sites with very high plasticity clays [$H > 25$ ft (7.6 m) and $PI > 75$] in a soil profile that would otherwise be classified as Site Class D or E, a site response analysis is not required provided both of the following requirements are satisfied: (i) Values of F_a and F_v are obtained from ASCE/SEI Tables 11.4-1 and 11.4-2, respectively, for Site Class D or E multiplied by the following factor: $$\begin{cases} 1.0 + \dfrac{0.3(PI - 75)}{50} & \text{for } 75 \leq PI \leq 125 \\ 1.3 & \text{for } PI > 125 \end{cases}$$ (ii) The resulting values of S_{DS} and S_{D1} obtained using the scaled factors F_a and F_v do not exceed the upper bound values for SDC B given in ASCE/SEI Tables 11.6-1 and 11.6-2. • For sites with very thick soft/medium stiff clays [$H > 120$ ft (36.6 m) and $s_u < 1,000$ lb/ft^2 (50 kN/m^2)], a site response analysis is not required provided both of the following requirements are satisfied: (i) Values of F_a and F_v are obtained from ASCE/SEI Tables 11.4-1 and 11.4-2, respectively, for Site Class E. (ii) The resulting values of S_{DS} and S_{D1} obtained using the factors F_a and F_v do not exceed the upper bound values for SDC B given in ASCE/SEI Tables 11.6-1 and 11.6-2.

	Ground motion hazard analysis in accordance with ASCE/SEI 21.2	
Seismically isolated structures and structures with damping systems on sites with $S_1 \geq 0.6$		—
Structures on Site Class E sites with $S_1 \geq 1.0$		A ground motion hazard analysis is not required for structures on Site Class E sites with $S_1 \geq 1.0$ provided F_a is taken as equal to that of Site Class C (that is, $F_a = 1.2$).
Structures on Site Class D and E sites with $S_1 \geq 0.2$		• A ground motion hazard analysis is not required for structures on Site Class D sites with $S_1 \geq 0.2$ provided the seismic response coefficient, C_s, is determined by ASCE/SEI Equation (12.8-2) for values of $T \leq 1.5T_s$ and is taken as equal to 1.5 times the value calculated in accordance with either ASCE/SEI Equation (12.8-3) for $1.5T_s < T \leq T_L$ or ASCE/SEI Equation (12.8-4) for $T > T_L$. • A ground motion hazard analysis is not required for structures on Site Class E sites with $S_1 \geq 0.2$ provided $T \leq T_s$ and the equivalent lateral force (ELF) procedure in ASCE/SEI 12.8 is used for design.

TABLE 2.8 Site-Specific Ground Motion Procedures

Additional information on conducting a site response analysis is given in ASCE/SEI C21.1 and in Ref. 8. A list of frequently used computer programs that can perform such an analysis is given in ASCE/SEI C21.1.3.

Risk-Targeted MCE$_R$ Ground Motion Hazard Analysis

Site-specific MCE_R ground motions are determined in accordance with ASCE/SEI 21.2 based on site-specific probabilistic and site-specific deterministic ground motions, both of which are defined in terms of the 5-percent damped spectral response in the maximum direction of the horizontal response. The results of a site-specific MCE_R ground motion hazard analysis are used in determining the MCE_R response spectrum (ASCE/SEI 21.2), the design response spectrum (ASCE/SEI 21.3), and the site-specific design accelerations S_{MS}, S_{M1}, S_{DS}, and S_{D1} (ASCE/SEI 21.4).

Probabilistic (MCE$_R$) Ground Motions

Two methods are provided in ASCE/SEI 21.2.1 on how to determine the spectral response acceleration, S_a. In Method 1, values of S_a are calculated for periods, T, by multiplying the risk coefficient, C_R, by the spectral response acceleration from a 5-percent damped acceleration response spectrum having a 2-percent probability of exceedance within a 50-year period. Values of C_{RS} and C_{R1}, which are used in determining C_R, are given in ASCE/SEI Figures 22-18 and 22-19, respectively:

- For $T \leq 0.2\ s: C_R = C_{RS}$
- For $T \geq 1.0\ s: C_R = C_{R1}$
- For $0.2\ s < T < 1.0\ s: C_R$ is based on linear interpolation of C_{RS} and C_{R1}

In lieu of using ASCE/SEI Figures 22-18 and 22-19, values of C_{RS} and C_{R1} can be obtained from Refs. 6 and 7 for a given site.

In Method 2, values of S_a are computed from iterative integration of a site-specific hazard curve with a lognormal probability density function representing the probability of collapse. A 1-percent probability of collapse within a 50-year period must be achieved (ASCE/SEI 21.2.1.2).

Deterministic (MCE$_R$) Ground Motions

Deterministic MCE_R ground motions are determined in accordance with ASCE/SEI 21.2.2. Values of S_a at each period are calculated as an 84th-percentile, 5-percent damped spectral response acceleration in the direction of maximum horizontal response. In cases where the largest spectral response acceleration of the resulting deterministic ground motion response spectrum is less than $1.5F_a$, the response spectrum must be scaled by a single factor such that the maximum response spectral acceleration is equal to $1.5F_a$ (see Supplement 1). The value of F_a is determined as follows:

- For site classes A, B, C, and D, F_a must be determined using ASCE/SEI Table 11.4.1 with $S_S = 1.5$.
- For site class E, $F_a = 1.0$.

In cases where the largest spectral response acceleration of the probabilistic ground motion response spectrum determined by ASCE/SEI 21.2.1 is less than $1.2F_a$, the deterministic ground motion response spectrum need not be calculated.

Additional information on this method is given in Ref. 8.

Site-Specific MCE$_R$

Once the values of the S_a are determined from the probabilistic and deterministic methods, the site-specific MCE_R spectral response acceleration, S_{aM}, is set equal to the lesser of the values obtained from these two methods for a given period (ASCE/SEI 21.2.3). However, where the largest spectral response acceleration of the probabilistic ground motion response spectrum of ASCE/SEI 21.2.1 is less than $1.2F_a$, the site-specific MCE_R ground motion response spectrum must be taken as the probabilistic ground motion response spectrum of ASCE/SEI 21.2.1 (see Supplement 1). In such cases, the value of F_a is determined as follows:

- For site classes A, B, C, and D, F_a must be determined using ASCE/SEI Table 11.4.1 with $S_S = 1.5$.
- For site class E, $F_a = 1.0$.

The site-specific MCE_R spectral response acceleration at any period must not be taken less than 150 percent of the site-specific design response spectrum determined in accordance with ASCE/SEI 21.3.

Design Response Spectrum

The design spectral response acceleration, S_a, is determined by ASCE/SEI Equation (21.3-1):

$$S_a = \frac{2}{3} S_{aM} \tag{2.5}$$

where S_{aM} is obtained from ASCE/SEI 21.1 or 21.2.

The design spectral response acceleration at any period must not be taken less than 80 percent of S_a determined in accordance with ASCE/SEI 11.4.6 where F_a and F_v are determined as follows:

- For Site Class A, B, and C: F_a and F_v are determined using ASCE/SEI Tables 11.4-1 and 11.4-2, respectively.
- For Site Class D: F_a is determined using ASCE/SEI Table 11.4-1, $F_v = 2.4$ where $S_1 < 0.2$, and $F_v = 2.5$ where $S_1 \geq 0.2$.
- For Site Class E: F_a is determined using ASCE/SEI Table 11.4-1 where $S_S < 1.0$, $F_a = 1.0$ where $S_S \geq 1.0$, $F_v = 4.2$ where $S_1 \leq 0.1$, and $F_v = 4.0$ where $S_1 > 0.1$.

For sites classified as Site Class F requiring a site response analysis in accordance with ASCE/SEI 11.4.8, the design spectral response acceleration at any period must be greater than or equal to 80 percent of S_a determined for Site Class E. A lower limit of 80 percent of S_a is permitted to be used where a different site class can be justified using the site-specific classification procedures in accordance with ASCE/SEI 20.3.3.

Design Acceleration Parameters

Requirements on how to determine the design accelerations S_{DS} and S_{D1} where the site-specific procedure is used to determine the design ground motion in accordance with ASCE/SEI 21.3 are as follows (ASCE/SEI 21.4):

- $S_{DS} = 0.9S_a$ at any T within the range of 0.2 to 5 s, inclusive.
- S_{D1} = maximum value of TS_a for periods from 1 to 2 s for sites with $v_{s,30} > 1{,}200$ ft/s (365.8 m/s) and for periods from 1 to 5 s for sites with $v_{s,30} \leq 1{,}200$ ft/s (365.8 m/s).

The parameters S_{MS} and S_{M1} must be taken as 1.5 times S_{DS} and S_{D1}, respectively.

Values of S_{MS} and S_{M1} determined by this method must be taken greater than or equal to 80 percent of the values determined in accordance with ASCE/SEI 11.4.3 for S_{MS} and S_{M1}:

- $S_{MS} = 1.5S_{DS} \geq 0.8S_{MS}$ determined from ASCE/SEI 11.4.3
- $S_{M1} = 1.5S_{D1} \geq 0.8S_{M1}$ determined from ASCE/SEI 11.4.3

Similarly, S_{DS} and S_{D1} determined by this method must be taken greater than or equal to 80 percent of the values determined by ASCE/SEI 11.4.5 for S_{DS} and S_{D1}.

Where a site-specific ground motion procedure is used in conjunction with the equivalent lateral force (ELF) procedure of ASCE/SEI 12.8, S_a at T is permitted to replace S_{D1}/T in ASCE/SEI Equation (12.8-3) and $S_{D1}T_L/T^2$ in ASCE/SEI Equation (12.8-4). The following are also relevant:

- S_{DS} determined in accordance with ASCE/SEI 21.4 is permitted to be used in ASCE/SEI Equations (12.8-2), (12.8-5), (15.4-1), and (15.4-3).
- The mapped value of S_1 must be used in ASCE/SEI Equations (12.8-6), (15.4-2), and (15.4-4).

Maximum Considered Earthquake Geometric Mean (MCE_G) Peak Ground Acceleration

According to ASCE/SEI 21.5.3, the site-specific MCE_G peak ground acceleration, PGA_M, must be taken as the lesser of the following:

1. the probabilistic geometric mean peak ground acceleration of ASCE/SEI 21.5.1 (which is the geometric mean peak ground acceleration with a 2-percent probability of exceedance within a 50-year period) and
2. the deterministic geometric mean peak ground acceleration of ASCE/SEI 21.5.2 (which is the largest 84th-percentile geometric mean peak ground acceleration for characteristic earthquakes on all known active faults within the site region).

The deterministic geometric mean peak ground acceleration must be greater than or equal to $0.5F_{PGA}$ where the site coefficient F_{PGA} is determined using ASCE/SEI Table 11.8-1 with the value of PGA taken as $0.5g$ (ASCE/SEI 21.5.2).

The site-specific MCE_G peak ground acceleration adjusted for site class effects, PGA_M, which is used for evaluation of liquefaction, lateral soil movement, seismic settlements, and other soil-related issues (see ASCE/SEI 11.8.3), must be greater than or equal to 80 percent of PGA_M determined by ASCE/SEI Equation (11.8-1) (ASCE/SEI 21.5.3).

2.3   Importance Factor and Risk Category

Risk categories are defined in IBC Table 1604.5 and ASCE/SEI Table 1.5-1. These categories are used to relate the criteria for maximum environmental loads or distortions specified in the code or referenced standards to the consequence that would occur to the structure and its occupants if such loads were exceeded.

A seismic importance factor, I_e, is assigned to a building or structure in accordance with ASCE/SEI Table 1.5-2 based on its risk category (see Table 2.9). Larger values of I_e are assigned to more important risk categories, such as assembly and essential facilities, to increase the likelihood that such structures would suffer less damage and continue to

Risk Category	I_e
I, II	1.00
III	1.25
IV	1.50

TABLE 2.9 Importance Factor, I_e

function during and following a design earthquake. The risk category of a structure is also used in determining the SDC (see Sec. 2.4 of this publication).

Requirements pertaining to operational access to a Risk Category IV structure through an adjacent structure are given in ASCE/SEI 11.5.2. In such cases, the adjacent structure must conform to the requirements for Risk Category IV structures. Additionally, where operational access is less than 10 ft (3.1 m) from an interior lot line or another structure on the same lot, protection from potential falling debris from adjacent structures must be provided by the owner of the Risk Category IV structure.

2.4 Seismic Design Category

All buildings and structures must be assigned to an SDC in accordance with IBC 1613.2.5 or ASCE/SEI 11.6. In general, SDC is a function of the risk category and the design spectral accelerations at the site.

Six SDCs are defined ranging from A (minimal seismic risk) to F (highest seismic risk). As the SDC of a structure increases, so do the strength and detailing requirements.

The SDC of a building or structure is assigned as follows:

- For $S_1 \geq 0.75$: $\begin{cases} \text{Risk Category I, II, or III structures are assigned to SDC E} \\ \text{Risk Category IV structures are assigned to SDC F} \end{cases}$

- For $S_1 < 0.75$: $\begin{cases} \text{Determine the SDC as a function of } S_{DS} \text{ by ASCE/SEI Table 11.6.1} \\ \text{Determine the SDC as a function of } S_{D1} \text{ by ASCE/SEI Table 11.6.2} \end{cases}$

The more severe SDC of the two determined by ASCE/SEI Tables 11.6.1 and 11.6.2 governs.

The SDC may be determined by ASCE/SEI Table 11.6-1 based solely on S_{DS} in cases where $S_1 < 0.75$ provided all the following conditions in ASCE/SEI 11.6 are satisfied:

1. In each of the two orthogonal directions, the approximate fundamental period of the structure, T_a, determined in accordance with ASCE/SEI 12.8.2.1 is less than $0.8T_S = 0.8S_{D1}/S_{DS}$.

2. In each of the two orthogonal directions, the fundamental period of the structure, T, used to calculate the story drift is less than $T_S = S_{D1}/S_{DS}$.

3. ASCE/SEI Equation (12.8-2) is used to determine the seismic response coefficient C_S.

4. The diaphragms are rigid in accordance with ASCE/SEI 12.3.1; or, for diaphragms that are not rigid, the horizontal distance between vertical elements of the seismic force–resisting system (SFRS) does not exceed 40 ft (12.2 m).

This exception should always be considered—especially when determining the SDC of low-rise buildings that are stiff (that is, buildings with small structural periods)—because it is possible for a building to be assigned to a lower SDC. As noted above, a lower SDC generally means less stringent design and detailing requirements.

For structures that can be designed in accordance with the alternate simplified design procedure in ASCE/SEI 12.14, the SDC is permitted to be determined from ASCE/SEI Table 11.6-1 alone, using the value of S_{DS} determined in ASCE/SEI 12.14.8.1. However, in cases where $S_1 \geq 0.75$, the building or structure must be assigned to SDC E. The provisions for this method are given in Sec. 3.15 of this publication.

The SDC is a trigger mechanism for many seismic requirements, including the following:

- Permissible seismic force resisting systems.
- Limitations on building height.
- Consideration of structural irregularities.
- The need for additional special inspections, structural testing, and structural observation for seismic resistance.

The flowchart in Fig. 2.5 can be used to determine the SDC of a building or structure.

2.5 Design Requirements for SDC A

Structures assigned to SDC A need only comply with the general structural integrity requirements of ASCE/SEI 1.4. The lateral force–resisting system must be proportioned to resist a lateral force, F_x, at each floor level equal to 1 percent of the total dead load, W_x, located or assigned at that floor level (see Fig. 2.6).

According to ASCE/SEI 1.4.2, the lateral forces are to be applied independently in each of two orthogonal directions.

Requirements for load path connections, connection to supports, and anchorage of structural walls are given in ASCE/SEI 1.4.1, 1.4.3, and 1.4.4, respectively.

Nonstructural components of a building or structure assigned to SDC A are automatically exempt from any seismic design requirements (ASCE/SEI 11.7). Also, tanks assigned to Risk Category IV must satisfy the freeboard requirement in ASCE/SEI 15.6.5.1.

2.6 Geological Hazards and Geotechnical Investigation

2.6.1 Site Limitation for SDC E and F

Any structure assigned to SDC E or F must not be located where there is a potential for an active fault to cause rupture of the ground surface at that location (ASCE/SEI 11.8.1). It is very difficult to design structures to accommodate large movement associated with ground fault ruptures; thus, buildings are not permitted at these locations.

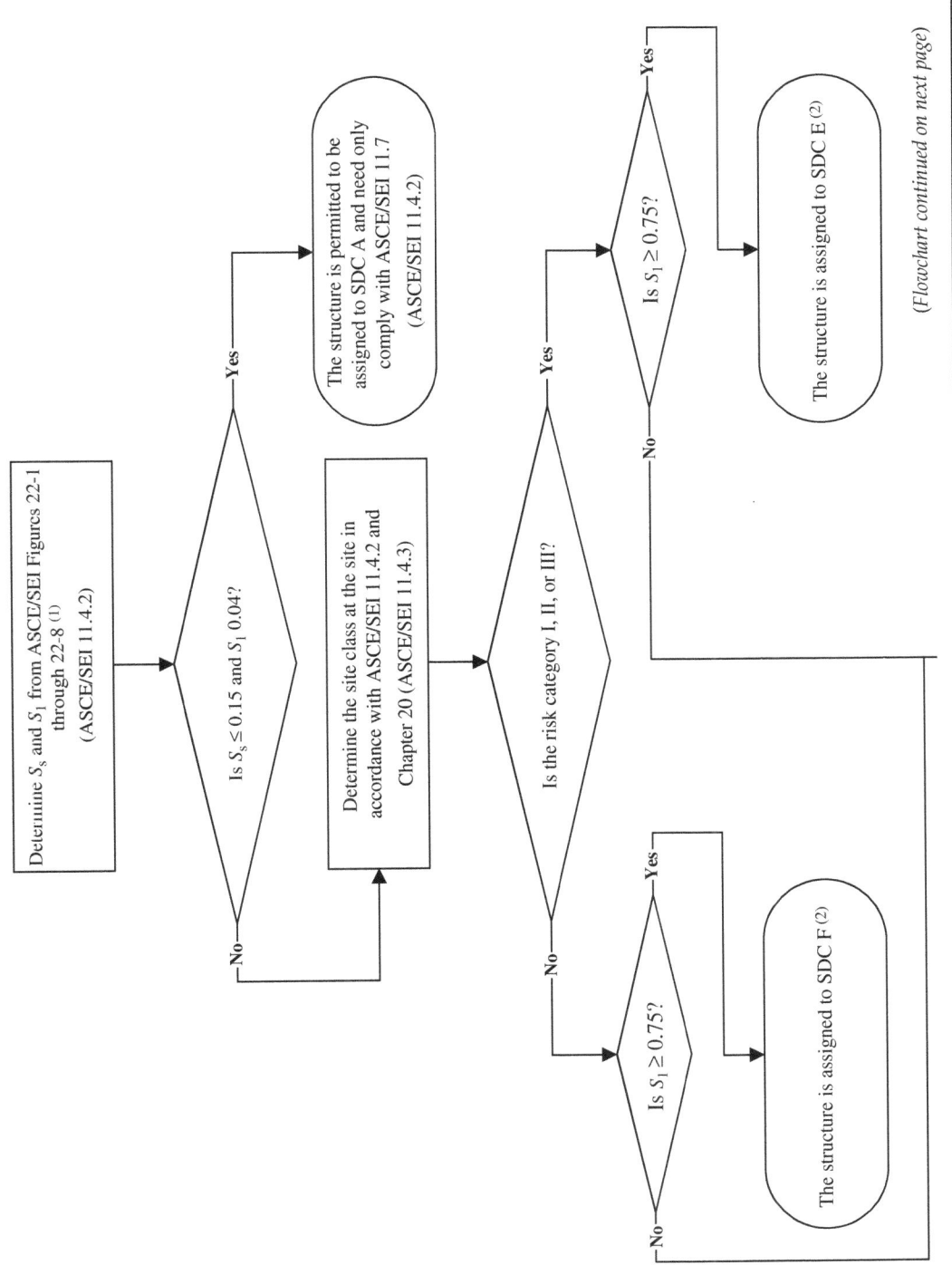

The structure is permitted to be assigned to SDC A and need only comply with ASCE/SEI 11.7 (ASCE/SEI 11.4.2)

Determine S_s and S_1 from ASCE/SEI Figures 22-1 through 22-8 [1] (ASCE/SEI 11.4.2)

Is $S_s \leq 0.15$ and S_1 0.04?

Determine the site class at the site in accordance with ASCE/SEI 11.4.2 and Chapter 20 (ASCE/SEI 11.4.3)

Is the risk category I, II, or III?

Is $S_1 \geq 0.75$?

The structure is assigned to SDC E [2]

Is $S_1 \geq 0.75$?

The structure is assigned to SDC F [2]

Yes

No

Yes

No

Yes

No

Yes

No

(Flowchart continued on next page)

FIGURE 2.5 Determination of SDC in accordance with ASCE/SEI 11.6.

31

32

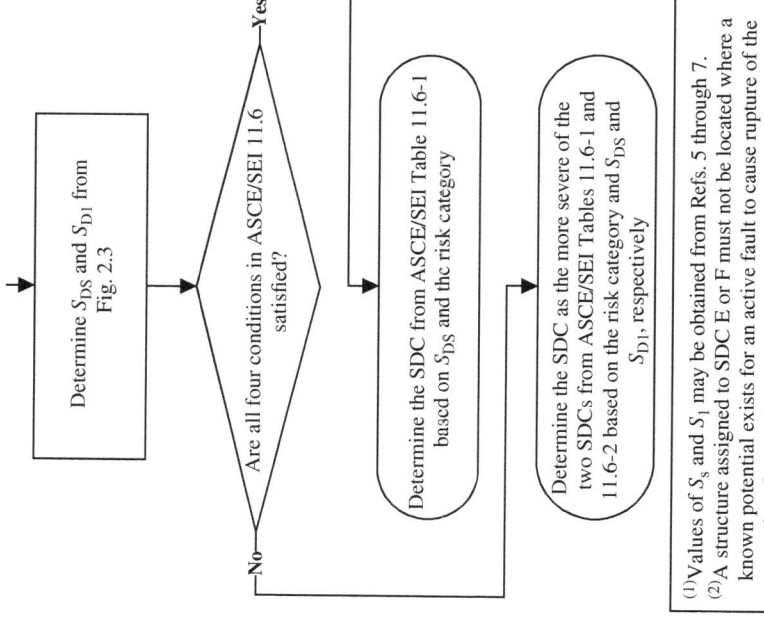

Determine S_{DS} and S_{D1} from Fig. 2.3

Are all four conditions in ASCE/SEI 11.6 satisfied?

Yes

No

Determine the SDC from ASCE/SEI Table 11.6-1 based on S_{DS} and the risk category

Determine the SDC as the more severe of the two SDCs from ASCE/SEI Tables 11.6-1 and 11.6-2 based on the risk category and S_{DS} and S_{D1}, respectively

[1]Values of S_s and S_1 may be obtained from Refs. 5 through 7.
[2]A structure assigned to SDC E or F must not be located where a known potential exists for an active fault to cause rupture of the ground surface at the structure (ASCE/SEI 11.8.1).

FIGURE 2.5 (Continued)

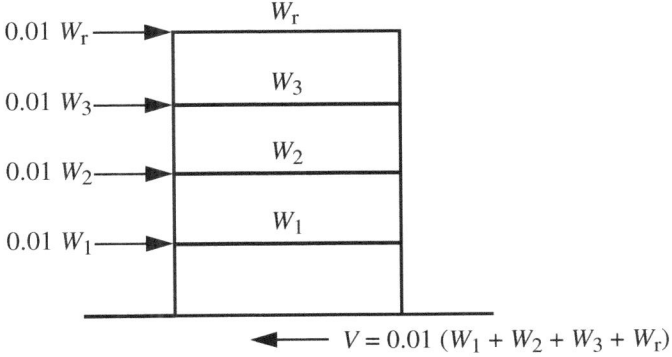

$$V = 0.01\ (W_1 + W_2 + W_3 + W_r)$$

Figure 2.6 General structural requirements of ASCE/SEI 1.4 for structures assigned to SDC A.

2.6.2 Geotechnical Investigation Report Requirements for Seismic Design Categories C through F

For structures assigned to SDC C, D, E, or F, the geotechnical report for a particular site must include an evaluation of the following potential geologic and seismic hazards:

- Slope instability
- Liquefaction
- Total and differential settlement
- Surface displacement due to faulting or seismically induced lateral spreading or lateral flow

Recommendations on the type of foundation and other methods to mitigate any of the aforementioned hazards must also be included in the geotechnical report.

The exception in ASCE/SEI 11.8.2 states that a site-specific geotechnical report is not required at locations where evaluations of nearby sites with similar soil conditions have been performed previously; in such cases, the nearby information can be utilized for the proposed site provided the authority having jurisdiction approves it.

2.6.3 Additional Geotechnical Investigation Report Requirements for Seismic Design Categories D through F

For structures assigned to SDC D, E, or F, the geotechnical report must include the information in ASCE/SEI 11.8.3 in addition to the information discussed above (see Table 2.10).

It is essential to have a clear plan on how to control the potential effects from liquefaction and soil strength loss. Appropriate foundation and lateral force–resisting systems must be established in the early design stages to counteract any potential damaging effects.

Requirement	Remarks
The determination of dynamic seismic lateral earth pressures on basement and retaining walls due to design earthquake ground motions.	Dynamic lateral pressures are considered to be an earthquake load, E, which is superimposed on the preexisting lateral earth load, H, during ground shaking.
The potential for liquefaction and soil strength loss evaluated for site peak ground acceleration, earthquake magnitude, and source characteristics consistent with the maximum considered earthquake geometric mean (MCE_G) peak ground acceleration.	Peak accelerations are to be determined using one of the following methods: (a) a site-specific study that accounts for soil amplification effects as prescribed in the site-specific ground motion procedures of ASCE/SEI 11.4.8 (see Sec. 2.2.8 of this publication) or (b) the peak ground acceleration PGA_M determined by ASCE/SEI Equation (11.8-1): $$PGA_M = F_{PGA}PGA^*$$
Assessment of potential consequences of liquefaction and soil strength loss, including, but not limited to the following: (a) estimation of total and differential settlement (b) lateral soil movement (c) lateral soil loads on foundations (d) reduction in soil-bearing capacity and lateral soil reactions (e) soil downdrag and reduction in axial and lateral soil reaction for pile foundations (f) increases in soil lateral pressures on retaining walls (g) flotation of buried structures	Because the effects from liquefaction and soil strength loss can be disastrous, it is very important to have a clear understanding of how the effects will impact the structure and its foundation.
Discussion of mitigation measures, such as, but not limited to the following: (a) selection of appropriate foundation type and depth (b) selection of appropriate structural systems to accommodate anticipated displacements and forces (c) ground stabilization (d) any combinations of these measures and how they must be considered in the design of the structure	—

*In this equation, F_{PGA} is the site coefficient from ASCE/SEI Table 11.8-1 and PGA is the mapped MCE_G peak ground acceleration given in ASCE/SEI Figures 22-9 through 22-13 (PGA can also be obtained from Refs. 5 through 7). In cases where Site Class D is selected as the default site class in accordance with ASCE/SEI 11.4.3, the value of F_{PGA} must be taken as greater than or equal to 1.2 (ASCE/SEI 11.8.3(2)).

TABLE 2.10 Additional Geotechnical Investigation Report Requirements for SDCs D through F

2.7 Vertical Ground Motions for Seismic Design

The requirements in ACI 11.9 may be used in lieu of the requirements in ASCE/SEI 12.4.2.2 for the effects of vertical design earthquake motions on a structure. These requirements are meant for the design of structures where a significant response to vertical ground motion is possible (such as many types of nonbuilding structures). ASCE/SEI Chapter 15, Seismic Design Requirements for Nonbuilding Structures, is the only chapter that references these requirements. Additional information and an illustrative example of a vertical response spectrum can be found in ASCE/SEI C11.9 (also see Example 2.2 of this publication). Plots of the MCE_R and design vertical response spectra are part of the output from Ref. 5.

2.8 Examples

The following examples illustrate the determination of site classification, design spectral acceleration parameters, and SDC.

2.8.1 Example 2.1—Determination of Site Classification

Determine the site class of a commercial building using the requirements in ASCE/SEI Chapter 20. The site is not located in close proximity to a known fault and the soil profile over the top 100 ft (30.5 m) of the site is given in Table 2.11. The soil properties in Table 2.11 are based on a composite of the properties determined from several borings taken at the site.

Spoon refusal occurs 61 ft (18.6 m) from the ground surface. The maximum blow count of 100 blows/ft (305 blows/m) is assigned at that depth (ASCE/SEI 20.4.2). This blow count is used from refusal to a depth of 100 ft (30.5 m) to complete the site profile.

Solution

The flowchart in Fig. 2.2 is used to determine the site class.

Step 1—Determine if the site can be classified as Site Class F   ASCE/SEI 20.3.1

It is evident that none of the conditions in ASCE/SEI 20.3.1 are satisfied, so the site cannot be classified as Site Class F.

Step 2—Determine if the site can be classified as Site Class E   ASCE/SEI 20.3.2

Check if layer 3 is a soft clay layer (see ASCE/SEI Table 20.3-1):

- Layer thickness $d_i = 4$ ft (1.2 m) < 10 ft (3.1 m)
- Blow count $N_i = 3$ blows/ft < 15 blows/ft
- Plasticity index $PI = 35 > 20$
- Undrained shear strength $s_u = 400$ lb/ft² (19.2 kN/m²) < 500 lb/ft² (24.0 kN/m²)

The site cannot be classified as Site Class E because the thickness of layer 3 is less than 10 ft (3.1 m).

It is evident that layer 1 is also not a soft clay layer.

Step 3—Determine if the site can be classified as Site Class A or B
     ASCE/SEI 20.3.4 and 20.3.5

Layer I	Soil Designation[1]	d_i, ft (m)	Distance from Ground Surface to Bottom of Soil Layer, ft (m)	N_i	PI	w (%)	S_u, lb/ft² (kN/m²)	Soil Type[2]
1	Lean clay (CL)	3 (0.91)	3 (0.91)	7	24	15	1,350 (64.6)	Cohesive
2	Silty sand (ML)	12 (3.7)	15 (4.6)	8	—	—	—	Cohesionless
3	Fat clay (CH)	4 (1.2)	19 (5.8)	3	35	42	400 (19.2)	Cohesive
4	Sandy silt (ML)	10 (3.1)	29 (8.8)	7	—	—	—	Cohesionless
5	Poorly graded sand with silt (SP-SM)	22 (6.7)	51 (15.6)	38	—	—	—	Cohesionless
6	Poorly graded sand with gravel (SP)	10 (3.1)	61 (18.7)	47	—	—	—	Cohesionless
7	Rock	39 (11.9)	100 (30.5)	100	—	—	—	Cohesionless

(1) Soil classifications are based on the unified soil classification system.
(2) A cohesive soil layer is one in which $PI > 20$. A cohesionless soil layer is one in which $PI < 20$ (ASCE/SEI 20.3.3).

TABLE 2.11   Subsurface Soil Profile for the Commercial Building in Example 2.1

Layer	d_i, ft (m)	N_i	d_i/N_i
1	3 (0.91)	7	0.43
2	12 (3.7)	8	1.50
3	4 (1.2)	3	1.33
4	10 (3.1)	7	1.43
5	22 (6.7)	38	0.58
6	10 (3.1)	47	0.21
7	39 (11.9)	100	0.39
		Σ	5.87

Table 2.12 Calculations to Determine $\overline{N}$ for the Commercial Building in Example 2.1

Rock is present at a distance of 61 ft (18.7 m) from the ground surface. Assuming the building will be supported on spread footings, the site cannot be classified as Site Class A or Site Class B because there is more than 10 ft (3.1 m) of soil between the rock surface and the bottom of the spread footings (ASCE/SEI 20.1).

Step 4—Determine if the site can be classified as Site Class C or D

ASCE/SEI 20.3.3

The average field standard penetration resistance, $\overline{N}$, determined by ASCE/ SEI Equation (20.4-2) is used to determine the site class (using this method automatically excludes Site Classes A and B, which are based on shear wave velocity). Calculations to determine $\overline{N}$ are given in Table 2.12.

- $\overline{N} = \sum_{i=1}^{7} d_i / \sum_{i=1}^{7} (d_i/N_i) = 100/5.87 = 17$ blows/ft

Because $\overline{N} = 17$ blows/ft is between 15 and 50 blows/ft, the site is classified as Site Class D—stiff soil (see ASCE/SEI Table 20.3-1).

2.8.2 Example 2.2—Determination of Design Spectral Acceleration Parameters

Determine the design spectral acceleration parameters for the commercial building in Example 2.1. The site is located in Nashville, TN, at the following coordinates:

Latitude = 36.135°, Longitude = −86.780°

Solution

The flowchart in Fig. 2.3 is used to determine the design spectral acceleration parameters S_{DS} and S_{D1}.

Step 1—Determine S_S and S_1

ASCE/SEI 11.4.2

The mapped MCE_R, 5-percent damped, spectral response acceleration parameters at short periods, S_S, and at a period of 1 s, S_1, are determined using Refs. 5, 6, or 7.

For the given latitude and longitude coordinates of the site: $S_S = 0.282$ and $S_1 = 0.141$.

Because $S_S > 0.15$ and $S_1 > 0.04$, the structure is not permitted to be assigned to SDC A.

Step 2—Determine the site class ASCE/SEI 11.4.3

From Example 2.1, the site class is Site Class D—stiff soil.

A site response analysis in accordance with ASCE/SEI 21.1 need not be performed to determine the design spectral acceleration parameters because the site is not classified as F. A ground motion hazard analysis in accordance with ASCE/SEI 21.2 need not be performed because of the following: (1) the site is not classified as Site Class E with $S_S \geq 1.0$ or $S_1 \geq 0.2$ and (2) the site is classified as Site Class D with $S_1 < 0.2$.

Step 3—Determine S_{MS} and S_{M1} ASCE/SEI 11.4.4

The MCE_R spectral response acceleration parameters at short periods, S_{MS}, and at a period of 1 s, S_{M1}, are determined by ASCE/SEI Equations (11.4-1) and (11.4-2), respectively:

$$S_{MS} = F_a S_S$$

$$S_{M1} = F_a S_1$$

Site coefficients F_a and F_v are determined by Tables 2.4 and 2.5 of this publication, respectively (or from Refs. 5 through 7):

- For Site Class D—stiff soil and $0.25 < S_S = 0.282 < 0.50$, $F_a = 1.574$ by linear interpolation.

- For Site Class D—stiff soil and $0.10 < S_1 = 0.141 < 0.20$, $F_v = 2.318$ by linear interpolation.

Therefore,

$$S_{MS} = 1.574 \times 0.282 = 0.444$$

$$S_{M1} = 2.318 \times 0.141 = 0.327$$

Step 4—Determine S_{DS} and S_{D1} ASCE/SEI 11.4.5

The design spectral acceleration at short periods, S_{DS}, and at a period of 1 s, S_{D1}, are determined by ASCE/SEI Equations (11.4-3) and (11.4-4), respectively:

$$S_{DS} = 2S_{MS}/3 = 2 \times 0.444/3 = 0.296$$

$$S_{D1} = 2S_{M1}/3 = 2 \times 0.327/3 = 0.218$$

The design horizontal response spectrum in accordance with ASCE/SEI 11.4.6 is given in Fig. 2.7. The MCE_R response spectrum, which is determined by multiplying the design response spectrum by 1.5 (ASCE/SEI 11.4.7) is given in Fig. 2.8. Both response spectrums are part of the output from Refs. 5 and 6.

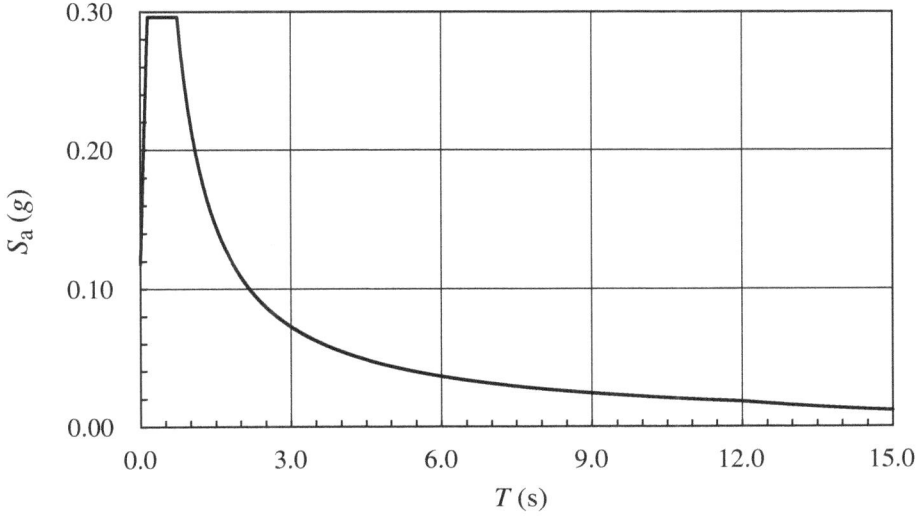

Figure 2.7 Design horizontal response spectrum for the commercial building in Example 2.2.

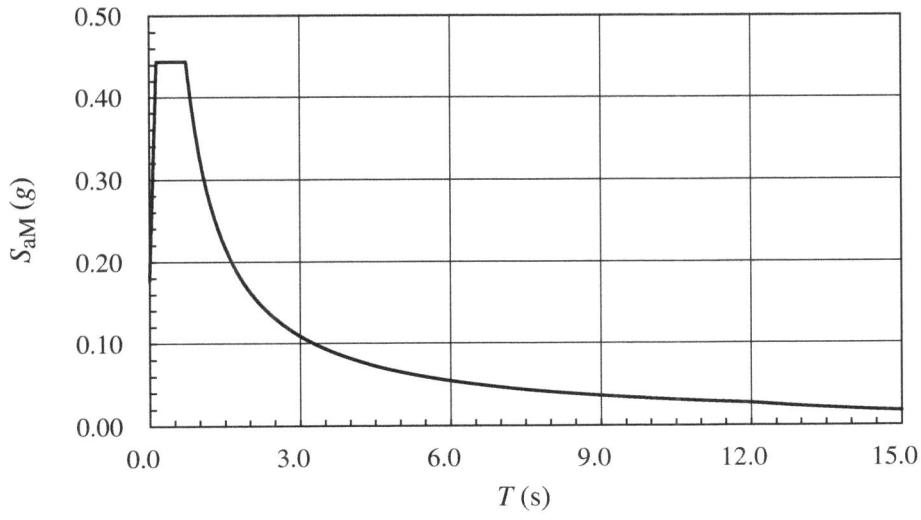

Figure 2.8 MCE_R response spectrum for the commercial building in Example 2.2.

MCE_R and design vertical response spectra are given in Figs. 2.9 and 2.10, respectively, for vertical periods of vibration, T_v. The equations in ASCE/SEI 11.9.2 to determine the MCE_R vertical response spectral accelerations, S_{aMv}, are given in Table 2.13 where values of the vertical coefficient, C_v, are obtained from ASCE/SEI Table 11.9-1. In accordance with ASCE/SEI 11.9.3, the design vertical response spectral accelerations, S_{av}, are taken as two-thirds of the

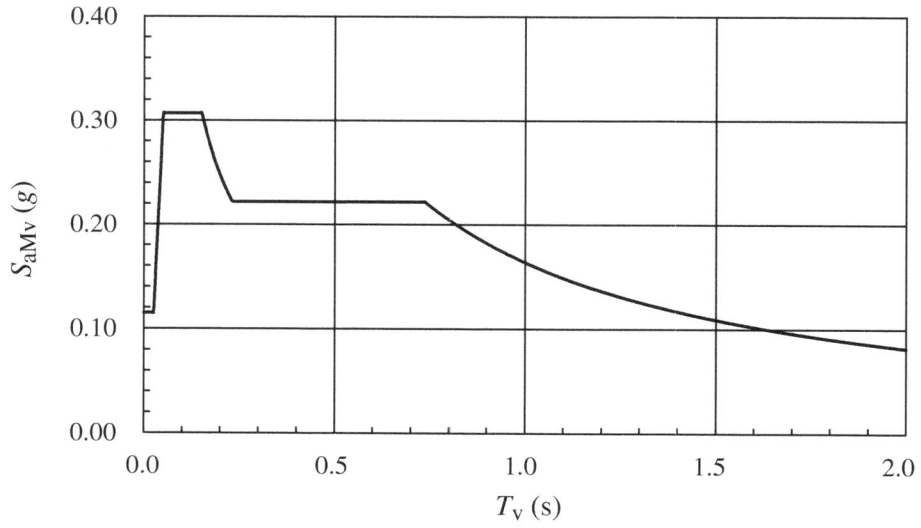

FIGURE 2.9 MCE_R vertical response spectrum for the commercial building in Example 2.2.

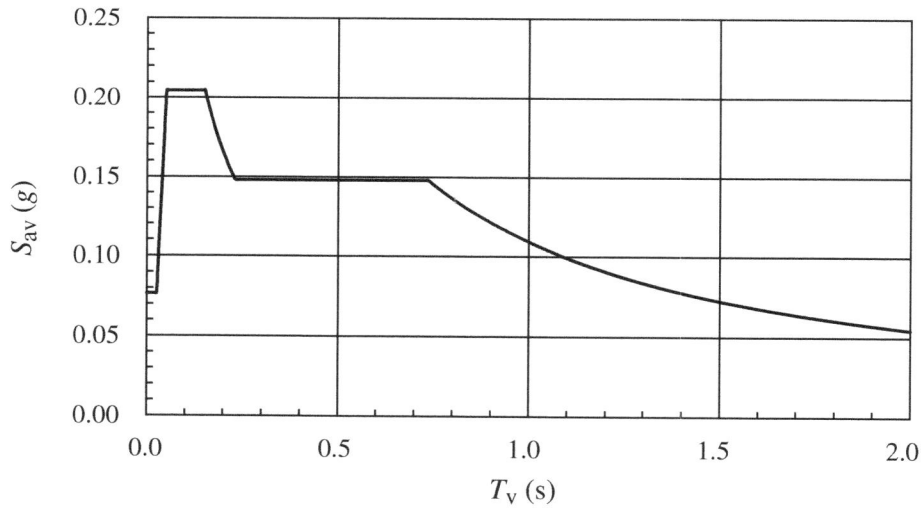

FIGURE 2.10   Design vertical response spectrum for the commercial building in Example 2.2.

value of S_{aMv}, where S_{aMv} must be taken equal to at least 50 percent of the corresponding MCE_R horizontal response spectral acceleration, S_{aM}.

2.8.3   Example 2.3—Determination of Seismic Design Category

Determine the SDC for the commercial building in Examples 2.1 and 2.2. Preliminary analysis indicates that the approximate period, T_a, is equal to 0.7 s.

ASCE/SEI Equation No.	Vertical Period, T_v	S_{aMv} (g)
11.9-1	$T_v \leq 0.025$ s	$S_{aMv} = 0.3C_vS_{MS} = 0.3 \times 0.864 \times 0.444 = 0.115\,^*$
11.9-2	0.025 s $< T_v \leq 0.05$ s	$S_{aMv} = 20C_vS_{MS}(T_v - 0.025) + 0.3C_vS_{MS}$
11.9-3	0.05 s $< T_v \leq 0.15$ s	$S_{aMv} = 0.8C_vS_{MS} = 0.307$
11.9-4	0.15 s $< T_v \leq 2.0$ s	$S_{aMv} = 0.8C_vS_{MS}(0.15/T_v)^{0.75}$

*Vertical coefficient, C_v, is determined by linear interpolation from ASCE/SEI Table 11.9-1.

TABLE 2.13 MCE_R Vertical Response Spectral Acceleration, S_{aMv}

Solution

The flowchart in Fig. 2.5 is used to determine the SDC.

Step 1—Determine S_S and S_1   ASCE/SEI 11.4.2

From step 1 in Example 2.2, $S_S = 0.282$ and $S_1 = 0.141$.

Step 2—Determine if the building can be assigned to SDC A   ASCE/SEI 11.4.2

Because $S_S = 0.282 > 0.15$ and $S_1 = 0.141 > 0.04$, the building cannot be assigned to SDC A.

Step 3—Determine the site class   ASCE/SEI 11.4.3

From Example 2.1, the site class is Site Class D—stiff soil.

Step 4—Determine the risk category   ASCE/SEI Table 1.5-1

For a commercial building, the risk category is II.

Step 5—Determine if the building is assigned to SDC E or F   ASCE/SEI 11.6

Because the risk category is II, the building is not assigned to SDC F.

Also, $S_1 = 0.141 < 0.75$, which means the building is not assigned to SDC E.

Step 6—Determine S_{DS} and S_{D1}   ASCE/SEI 11.4.5

From step 4 in Example 2.2, $S_{DS} = 0.296$ and $S_{D1} = 0.218$.

Step 7—Determine if the SDC can be determined using ASCE/SEI Table 11.6-1 alone

Check if all four conditions in ASCE/SEI 11.6 are satisfied:

- Check if $T_a < 0.8T_s$

$$0.8T_s = 0.8S_{D1}/S_{DS} = 0.8 \times 0.218/0.296 = 0.59 \text{ s} < T_a = 0.70 \text{ s}$$

Therefore, the SDC cannot be determined using ASCE/SEI Table 11.6-1 alone.

Step 8—Determine the SDC from ASCE/SEI Tables 11.6-1 and 11.6-2

From ASCE/SEI Table 11.6-1 with $S_{DS} = 0.296$ and Risk Category II, the SDC is B.

From ASCE/SEI Table 11.6-1 with $S_{D1} = 0.218$ and Risk Category II, the SDC is D.

Therefore, the SDC is D for this commercial building.

2.8.4   Example 2.4—Determination of a Site-Specific Design Response Spectrum

The results of a site response analysis in accordance with ASCE/SEI 21.1 are given in Table 2.14 for a residential building located on a Site Class D—stiff soil site where $S_S = 2.115$, $S_1 = 0.761$, and $T_L = 12$ s. The SDC has been determined to be E and the average shear wave velocity over a subsurface depth of 30 m (100 ft), $v_{s,30}$, has been determined to be 2,490 ft/s (759 m/s). Determine the site-specific design response spectrum in accordance with ASCE/SEI 21.3 and the design spectral accelerations parameters S_{DS}, S_{D1}, S_{MS}, and S_{M1} in accordance with ASCE/SEI 21.4.

Period (s)	Site-Specific MCE_R Response Spectrum, S_{aM}	Site-Specific Design Response Spectrum, $S_a = 2S_{aM}/3$ (ASCE/SEI 21.3)
0.01	0.94	0.63
0.02	1.08	0.72
0.03	1.23	0.82
0.05	1.52	1.01
0.065	1.74	1.16
0.075	1.88	1.25
0.085	2.01	1.34
0.10	2.01	1.34
0.15	2.01	1.34
0.20	2.01	1.34
0.30	2.34	1.56
0.40	2.64	1.76
0.50	2.31	1.54
0.75	1.83	1.22
0.90	1.67	1.11
1.00	1.61	1.07
1.50	1.12	0.75
2.00	0.86	0.57
3.00	0.49	0.33
4.00	0.36	0.24
5.00	0.29	0.19
7.50	0.18	0.12
10.00	0.13	0.09

TABLE 2.14   Site-Specific Horizontal MCE_R and Design Response Spectrums in Example 2.4

Solution

Step 1—Determine the site-specific design response spectrum   ASCE/SEI 21.3

The design site-specific spectral response acceleration at any period, T, must be greater than or equal to 80 percent of the design spectral response acceleration, S_a, determined in accordance with ASCE/SEI 11.4.6 where F_a and F_v are determined as follows for Site Class D (ASCE/SEI 21.3):

- F_a is determined using ASCE/SEI Table 11.4-1
- $F_v = 2.5$ for $S_1 = 0.761 > 0.2$

The equations to determine S_a based on T are given in Fig. 2.4.

A summary of the governing site-specific design response accelerations is given in Table 2.15. A comparison of the design response accelerations is given in Fig. 2.11.

Step 2—Determine S_{DS}, S_{D1}, S_{MS}, and S_{M1}   ASCE/SEI 21.4

1. S_{DS} = greater of the following:

   - 0.9 times the maximum S_a from the site-specific spectrum at any T within 0.2 to 5 s
   - $0.8 S_{DS}$ where S_{DS} is determined in accordance with ASCE/SEI 11.4.5

   The maximum S_a from the site-specific spectrum is equal to 1.76, which occurs at $T = 0.40$ s (see Table 2.15).

   Determine S_{DS} in accordance with ASCE/SEI 11.4.5 using the flowchart in Fig. 2.3:

   - Determine S_{MS} and S_{M1}

     $$F_a = 1.00 \text{ for } S_S = 2.115 > 1.5$$
     $$S_{MS} = F_a S_S = 1.00 \times 2.115 = 2.115 \qquad \text{ASCE/SEI Equation (11.4-1)}$$

     $$F_v = 1.70 \text{ for } S_1 = 0.761 > 0.6$$
     $$S_{M1} = F_v S_1 = 1.70 \times 0.761 = 1.294 \qquad \text{ASCE/SEI Equation (11.4-2)}$$

   - Determine S_{DS} and S_{D1}

     $$S_{DS} = 2S_{MS}/3 = 2 \times 2.115/3 = 1.410 \qquad \text{ASCE/SEI Equation (11.4-3)}$$
     $$S_{D1} = 2S_{M1}/3 = 2 \times 1.294/3 = 0.863 \qquad \text{ASCE/SEI Equation (11.4-4)}$$

   Therefore,

   $$S_{DS} = \text{greater of} \begin{cases} 0.9 \times 1.760 = 1.584 \text{ (governs)} \\ 0.8 \times 1.410 = 1.128 \end{cases}$$

Period (s)	Site-Specific MCE_R Response Spectrum, S_{aM}	Site-Specific Design Response Spectrum, $S_a = 2S_{aM}/3$ (ASCE/SEI 21.3)	S_a Determined in Accordance with ASCE/SEI 11.4.6	$0.8S_a$ Determined in Accordance with ASCE/ SEI 11.4.6	Governing Site-Specific Design Response Spectrum*	TS_a
0.01	0.94	0.63	0.61	0.49	0.63	0.006
0.02	1.08	0.72	0.66	0.53	0.72	0.014
0.03	1.23	0.82	0.71	0.57	0.82	0.025
0.05	1.52	1.01	0.80	0.64	1.01	0.051
0.065	1.74	1.16	0.87	0.70	1.16	0.075
0.075	1.88	1.25	0.92	0.74	1.25	0.094
0.085	2.01	1.34	0.96	0.77	1.34	0.114
0.10	2.01	1.34	1.03	0.82	1.34	0.134
0.15	2.01	1.34	1.27	1.02	1.34	0.201
0.20	2.01	1.34	1.41	1.13	1.34	0.268
0.30	2.34	1.56	1.41	1.13	1.56	0.468
0.40	2.64	1.76	1.41	1.13	1.76	0.704
0.50	2.31	1.54	1.41	1.13	1.54	0.770
0.75	1.83	1.22	1.41	1.13	1.22	0.915
0.90	1.67	1.11	1.41	1.13	1.13	1.017
1.00	1.61	1.07	1.27	1.02	1.07	1.070
1.50	1.12	0.75	0.85	0.68	0.75	1.125
2.00	0.86	0.57	0.63	0.50	0.57	1.140
3.00	0.49	0.33	0.42	0.34	0.34	1.020
4.00	0.36	0.24	0.32	0.26	0.26	1.040
5.00	0.29	0.19	0.25	0.20	0.20	1.000
7.50	0.18	0.12	0.17	0.14	0.14	1.050
10.00	0.13	0.09	0.13	0.10	0.10	1.000

*Larger of S_a determined by ASCE/SEI 21.3 and $0.8S_a$ determined by ASCE/SEI 11.4.6.

TABLE 2.15   Governing Site-Specific Design Response Accelerations in Example 2.4

2. For sites where $v_{s,30} > 1,200$ ft/s (365.8 m/s), S_{D1} = greater of the following:

- Maximum value of TS_a for periods from 1 to 2 s = 1.140 from Table 2.15 (governs)
- $0.8S_{D1}$ where S_{D1} is determined in accordance with ASCE/SEI 11.4.5 = $0.8 \times 0.863 = 0.690$

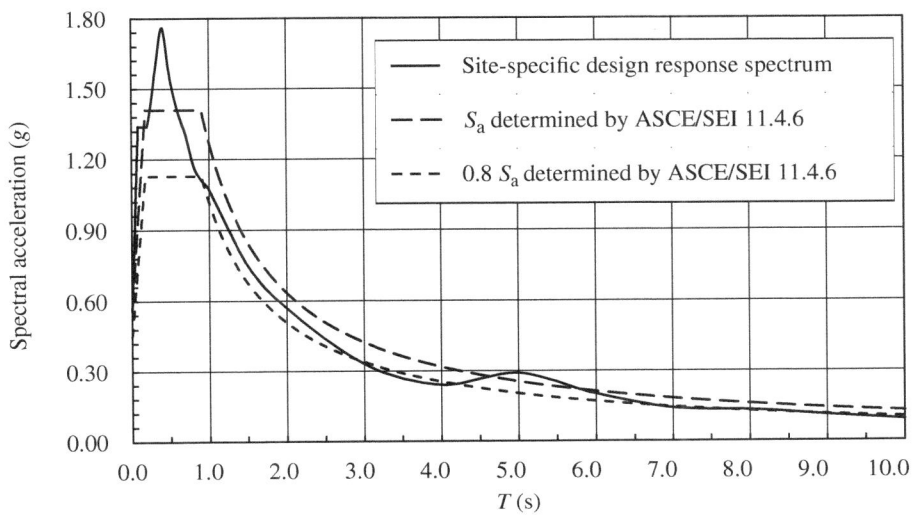

FIGURE 2.11 Design response accelerations in Example 2.4.

3. S_{MS} = greater of the following:

   - $1.5S_{DS}$ where S_{DS} is determined by ASCE/SEI 21.4 = $1.5 \times 1.584 = 2.376$ (governs)

   - $0.8S_{MS}$ where S_{MS} is determined in accordance with ASCE/SEI 11.4.5 = $0.8 \times 2.115 = 1.692$

4. S_{M1} = greater of the following:

   - $1.5S_{D1}$ where S_{D1} is determined by ASCE/SEI 21.4 = $1.5 \times 1.140 = 1.710$ (governs)

   - $0.8S_{M1}$ where S_{M1} is determined in accordance with ASCE/SEI 11.4.5 = $0.8 \times 1.294 = 1.035$

According to ASCE/SEI 21.4, the site-specific acceleration, S_a, at the structure period, T, is permitted to replace S_{D1}/T in ASCE/SEI Equation (12.8-3) and $S_{D1}/T_L/T^2$ in ASCE/SEI Equation (12.8-4) when calculating the design seismic base shear, V, in accordance with the ELF procedure. The design acceleration parameter, S_{DS}, determined in accordance with ASCE/SEI 21.4 is permitted to be used in ASCE/SEI Equations (12.8-2), (12.8-5), (15.4-1), and (15.4-3). The mapped value of S_1 must be used in ASCE/SEI Equations (12.8-6), (15.4-2), and (15.4-4).

CHAPTER 3

Seismic Design Requirements for Building Structures

3.1 Overview

Seismic analysis and design requirements applicable to the design of building structures and their members are given in this chapter, which are based on the provisions in ASCE/SEI Chapter 12. The following requirements are covered:

- Structural design basis (Sec. 3.2)
- Structural system selection (Sec. 3.3)
- Diaphragm flexibility, configuration irregularities, and redundancy (Sec. 3.4)
- Seismic load effects and combinations (Sec. 3.5)
- Direction of loading (Sec. 3.6)
- Analysis procedure selection (Sec. 3.7)
- Modeling criteria (Sec. 3.8)
- Equivalent lateral force procedure (Sec. 3.9)
- Linear dynamic analysis (Sec. 3.10)
- Diaphragms, chords, and collectors (Sec. 3.11)
- Structural walls and their anchorage (Sec. 3.12)
- Drift and deformation (Sec. 3.13)
- Foundation design (Sec. 3.14)
- Simplified alternative structural design criteria (Sec. 3.15)

3.2 Structural Design Basis

Basic requirements for seismic analysis and design of building structures are given in ASCE/SEI 12.1. A summary of these requirements is given in Table 3.1.

3.3 Structural System Selection

3.3.1 Selection and Limitations

The basic lateral and vertical SFRSs in a building structure must conform to (1) one of the types indicated in ASCE/SEI Table 12.2-1 or (2) a combination of the systems indicated

	Requirements	ASCE/SEI Section No.
Basic requirements	• The building structure must have a complete lateral and vertical force–resisting system capable of providing adequate strength, stiffness, and energy dissipation. • Design ground motions are assumed to occur along any horizontal direction. • The adequacy of the structural system must be demonstrated through the construction of a mathematical model and evaluation of the model for the effects of the design ground motions. • The design seismic forces must be determined using one of the applicable procedures indicated in ASCE/SEI 12.6.	12.1.1
Member design, connection design, and deformation limit	• Individual structural members must be designed to resist the effects due to the design ground motion. • Connections must develop the strength of the connected members or the forces indicated in ASCE/SEI 12.1.1. • The deformation of the structure subject to the design seismic forces must not exceed the prescribed limits.	12.1.2
Continuous load path and interconnection	• A continuous load path, or paths, with adequate strength and stiffness must be provided to transfer all forces from the point of application to the final point of resistance. • All parts of the structure between separation joints must be interconnected to form a continuous path to the seismic force–resisting system (SFRS). The connection must be capable of transmitting the seismic force F_p induced by the parts being connected. • Any smaller portion of a structure must be tied to the remainder of the structure with elements having a design strength capable of transmitting a seismic force equal to the greater of (1) $0.133S_{DS}$ times the weight of the smaller portion or (2) 0.05 times the portion's weight. This connection force does not apply to the overall design of the SFRS. • Connection design forces need not exceed the maximum forces that the structural system can deliver to the connection.	12.1.3
Connection to supports	• For each beam, girder, or truss in a structure, a positive connection must be provided for resisting a horizontal force acting parallel to the member, either directly to its supporting elements or to slabs designed as diaphragms. • The member's supporting element must be connected to the diaphragm in cases where the connection is through the diaphragm. • The connection must have a minimum design strength equal to 0.05 times the dead plus live load reaction.	12.1.4

TABLE 3.1 Basic Requirements for Seismic Analysis and Design of Building Structures

	Requirements	ASCE/SEI Section No.
Foundation design	• Foundations must be designed to resist the forces developed and to accommodate the movements imparted to the structure and foundation by the design ground motions. • The following must be included in the determination of the foundation design criteria: (1) dynamic natures of the forces, (2) expected ground motion, (3) design basis for the strength and energy dissipation capacity of the structure, and (4) the dynamic properties of the soil. • The design and construction of foundations must comply with ASCE/SEI 12.13. • The weights of foundations must be considered as dead loads in accordance with ASCE/SEI 3.1.2 when calculating load combinations using ASCE/SEI 2.3 or 2.4. The dead loads are permitted to include overlying fill and paving materials.	12.1.5

TABLE 3.1   Basic Requirements for Seismic Analysis and Design of Building Structures (*Continued*)

in ASCE/SEI Table 12.2-1 as permitted in ASCE/SEI 12.2.2, 12.2.3, and 12.2.4. The general categories of the SFRSs are given in Table 3.2.

The primary materials of construction are included under the general categories of the SFRSs in ASCE/SEI Table 12.2-1. The SFRSs are categorized according to the quality and extent of the seismic-resistant detailing that must be using in the design of the structure:

- "Special" systems provide superior seismic resistance, which is accomplished through extensive design and detailing of the structural members.
- "Ordinary" systems have basic design and detailing requirements.
- "Intermediate" systems provide seismic resistance that is better than ordinary systems, but not as good as special systems.

Design and detailing requirements for these three categories are given in the material standards for each type of material.

A summary of the seismic parameters in ASCE/SEI Table 12.2-1 is given Table 3.3.

The relationships between R, Ω_0, C_d are illustrated in Fig. 3.1, which is an inelastic force-deformation curve representative of a moment-resisting frame. The structure responds elastically until the first hinge forms. As the horizontal forces increase, a properly designed and detailed structure will respond inelastically with a series of plastic hinges forming due to the redundancy built into the SFRS. A yielding mechanism occurs at the strength level V_y, which is larger than the design seismic force, V (that is, $V_y = \Omega_0 V$).

Limitations on seismic design category and building height are also given in ASCE/SEI Table 12.2-1. Some SFRSs can be utilized in structures assigned to any SDC with no

SFRS	Description
Bearing wall system	A bearing wall system is a structural system where bearing walls support all or a major portion of the vertical loads. Some or all of the walls also provide resistance to the seismic forces.
Building frame system	In a building frame system, an essentially complete space frame provides support for the vertical loads. Shear walls or braced frames provide resistance to the seismic forces.
Moment-resisting frame system	A moment-resisting frame system is a structural system with an essentially complete space frame that supports the vertical loads. Seismic forces are resisted primarily by flexural action of the frame members through the joints. The entire space frame or selected portions of the frame may be designated as the SFRS. Requirements for deformation compatibility must be satisfied for structures assigned to SDC D, E, or F (ASCE/SEI 12.12.5).
Dual system	In a dual system, an essentially complete space frame provides support for the vertical loads. Moment-resisting frames and shear walls or braced frames provide resistance to seismic forces in accordance with their rigidities (see ASCE/SEI 12.2.5.1). Additionally, the moment-resisting frames must act as a backup for the walls or braces; this is accomplished by requiring that the moment frames be capable of resisting at least 25 percent of the design seismic forces. Dual systems are also referred to as shear wall-frame interactive systems.
Cantilever column system	In this system, the vertical forces and the seismic forces are resisted entirely by columns acting as cantilevers from their base. This system is usually used in one-story buildings or in the top story of a multistory building. Severe restrictions are placed on the use of this system because it has performed poorly in past earthquakes.

TABLE 3.2 General Categories of the SFRSs in ASCE/SEI Table 12.2-1

height limitations, while others are permitted in structures up to certain heights. The least ductile systems are not permitted in the higher SDCs regardless of height.

The use of SFRSs not in ASCE/SEI Table 12.2-1 is permitted provided the requirements in ASCE/SEI 12.2.1 are satisfied, including approval by the authority having jurisdiction.

3.3.2 Combinations of Framing Systems in Different Directions

It is permitted to use different SFRSs along each of the two orthogonal axes of a structure provided the respective R, Ω_0, and C_d values are used in each direction (ASCE/SEI 12.2.2).

It is possible for one of the systems to have more restrictive limitations on the structural system or the structural height. If this occurs, the more restrictive limitations govern for the entire building. Consider the building in Fig. 3.2, which is assigned to SDC D. In the north-south direction, the SFRS consists of special reinforced concrete moment frames at the two ends of the building (moment-resisting frame system C5 in ASCE/SEI Table 12.2-1) where $R = 8$, $\Omega_0 = 3$, and $C_d = 5.5$. In the east-west direction, the SFRS

Seismic Parameter	Description
Response modification coefficient, R	This coefficient accounts for the ability of an SFRS to respond to ground shaking in a ductile manner without loss of load-carrying capacity. In other words, R is an approximate way of accounting for the effective damping and energy dissipation that can be mobilized during inelastic response to ground shaking. It represents the ratio of the forces that would develop under the ground motion specified in ASCE/SEI 7 if the structure had responded to the prescribed design forces in an entirely linear-elastic manner.
	A system that has no ability to respond in a ductile manner has an R-value equal to 1; the only such system is a cantilevered column system consisting of ordinary reinforced concrete moment frames (see ASCE/SEI Table 12.2-1). Systems capable of highly ductile response have an R-value equal to 8.
Overstrength factor, Ω_0	The purpose of this factor is to amplify the prescribed seismic forces for use in design of the following types of structural elements: (1) elements whose action cannot provide reliable inelastic response or energy dissipation and (2) elements required to remain essentially elastic to maintain the structural integrity of the structure.
	For most of the structural systems given in the table, Ω_0 ranges from 2 to 3.
Deflection amplification factor, C_d	This factor is used to increase the elastic lateral displacements, δ_{xe}, determined for a structure using the prescribed design seismic forces to the lateral displacements expected during the design earthquake ground motion.
	It is evident from ASCE/SEI Table 12.2-1 that C_d is equal to or slightly less than the corresponding R-value for a given SFRS. The more ductile a system is (that is, the greater the R-value), the greater the difference is between the values of R and C_d.

TABLE 3.3   Seismic Parameters in ASCE/SEI Table 12.2-1

consists of special reinforced concrete shear walls (building frame system B4 in ASCE/ SEI Table 12.2-1) where $R = 6$, $\Omega_0 = 2.5$, and $C_d = 5$. The respective R, Ω_0, and C_d values are used in each direction. The building height is limited to 160 ft (48.8 m) based on the structural height limit of the special reinforced concrete shear wall system (there is no height limit for the special reinforced concrete moment frame system; see ASCE/SEI Table 12.2-1).

3.3.3   Combinations of Framing Systems in the Same Direction

Where different SFRSs are used in the same direction to resist seismic forces (other than combinations considered as dual systems), the most stringent structural system limitations in ASCE/SEI Table 12.2-1 apply. The intent of these requirements is to prevent concentration of inelastic behavior in the lower stories of a structure.

Vertical Combinations of SFRSs

Requirements for structures with a vertical combination of SFRSs in the same direction are given in ASCE/SEI 12.2.3.1 (see Fig. 3.3).

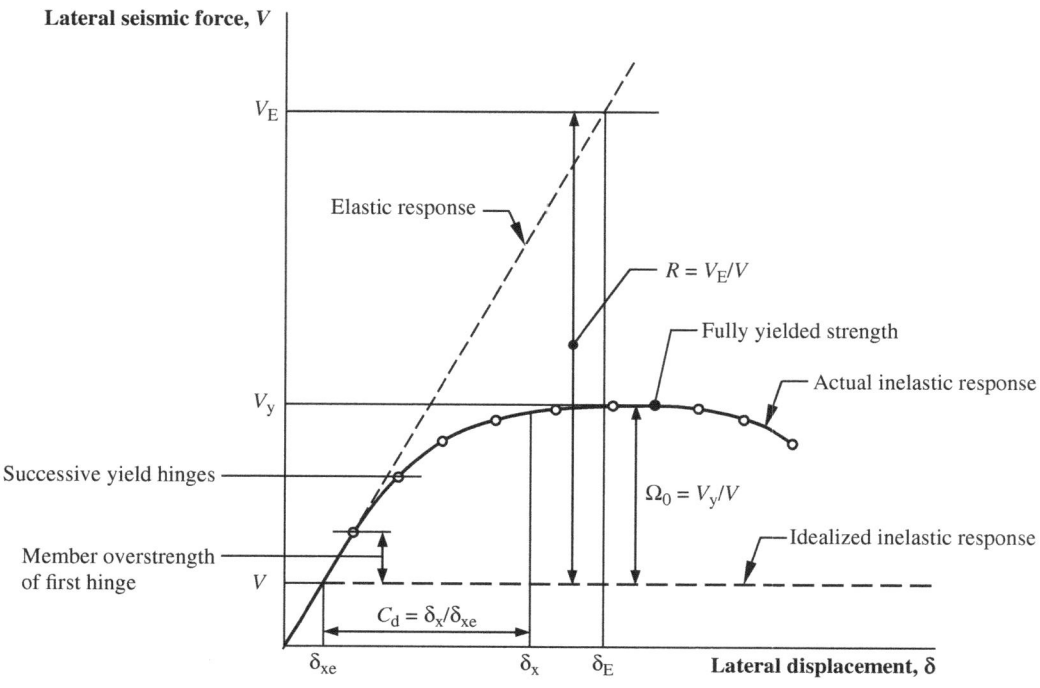

FIGURE 3.1 Inelastic force-deformation curve.

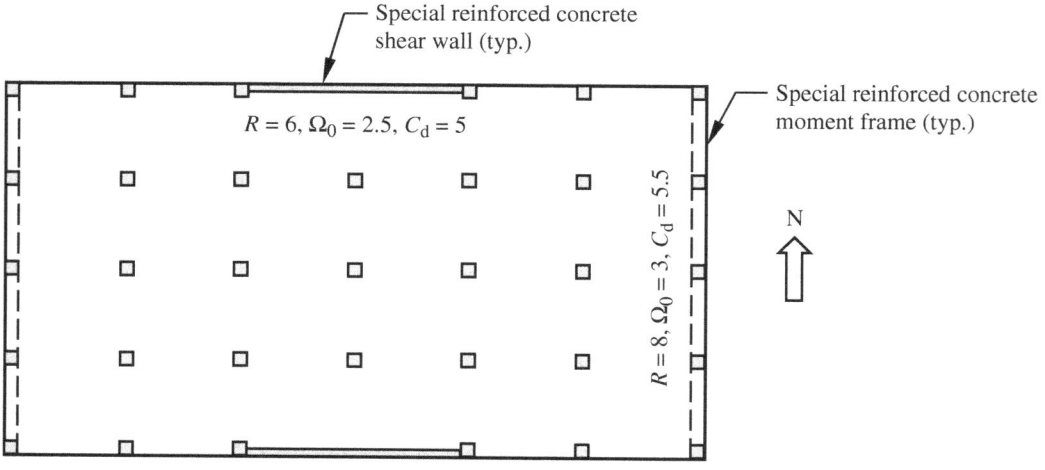

FIGURE 3.2 Combinations of framing systems in different directions.

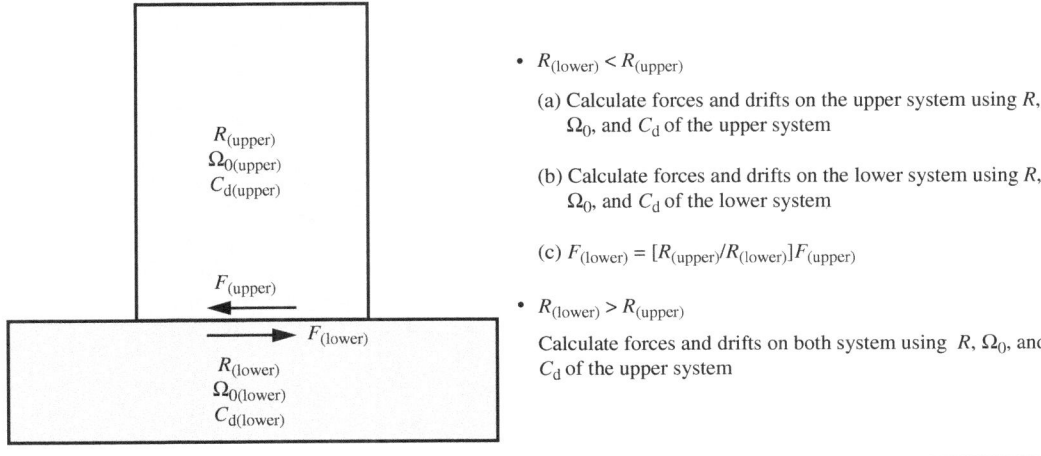

- $R_{(lower)} < R_{(upper)}$

  (a) Calculate forces and drifts on the upper system using R, Ω_0, and C_d of the upper system

  (b) Calculate forces and drifts on the lower system using R, Ω_0, and C_d of the lower system

  (c) $F_{(lower)} = [R_{(upper)}/R_{(lower)}]F_{(upper)}$

- $R_{(lower)} > R_{(upper)}$

  Calculate forces and drifts on both system using R, Ω_0, and C_d of the upper system

FIGURE 3.3 Vertical combinations of SFRSs.

Three exceptions to these requirements are given in ASCE/SEI 12.2.3.1:

1. Rooftop structures less than or equal to two stories in height and weighing less than 10 percent of the total structure weight.

2. Other supported structural systems with a weight less than or equal to 10 percent of the weight of the structure.

3. Detached one- and two-family dwellings of light-frame construction.

Two-Stage Analysis Procedure
A two-stage analysis is permitted in buildings with a flexible upper portion supported by a rigid lower portion provided the conditions in ASCE/SEI 12.2.3.2 are satisfied (see Fig. 3.4). An example of where this analysis procedure may be used is a reinforced concrete podium supporting a structure of light-frame construction.

Horizontal Combinations of SFRSs
For structures utilizing different SFRSs in the same horizontal direction, the least value of R and the corresponding C_d and Ω_0 are to be used in design in that direction (ASCE/SEI 12.2.3.3). The SFRSs for the building in Fig. 3.5 in the north-south direction consist of ordinary reinforced concrete shear walls ($R = 5, \Omega_0 = 2.5, C_d = 4.5$) and ordinary reinforced concrete moment frames ($R = 3, \Omega_0 = 3, C_d = 2.5$). Assuming a rigid diaphragm, the structure must be designed using $R = 3, \Omega_0 = 3$, and $C_d = 2.5$ in this direction (note: in lieu of utilizing separate SFRSs in this direction, the shear wall-frame interactive system with ordinary reinforced concrete moment frames and ordinary reinforced concrete shear walls in ASCE/SEI 12.2-1 may be used with $R = 4.5, \Omega_0 = 2.5$, and $C_d = 4$ provided all the appropriate requirements are satisfied).

According to the exception in ASCE/SEI 12.2.3.3, Risk Category I or II buildings two stories or less in height with light-frame construction or flexible diaphragms are permitted to be designed using the least value of R for the different SFRSs in each

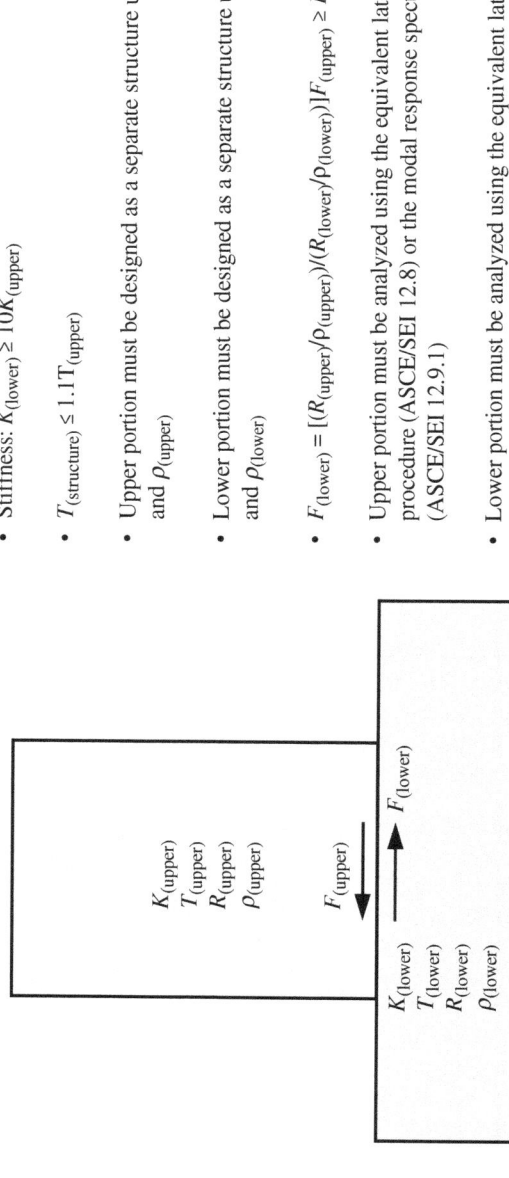

- Stiffness: $K_{(lower)} \geq 10K_{(upper)}$

- $T_{(structure)} \leq 1.1T_{(upper)}$

- Upper portion must be designed as a separate structure using $R_{(upper)}$ and $\rho_{(upper)}$

- Lower portion must be designed as a separate structure using $R_{(lower)}$ and $\rho_{(lower)}$

- $F_{(lower)} = [(R_{(upper)}/\rho_{(upper)})/(R_{(lower)}/\rho_{(lower)})]F_{(upper)} \geq F_{(upper)}$

- Upper portion must be analyzed using the equivalent lateral force procedure (ASCE/SEI 12.8) or the modal response spectrum analysis (ASCE/SEI 12.9.1)

- Lower portion must be analyzed using the equivalent lateral force procedure (ASCE/SEI 12.8)

Figure 3.4 Requirements for a two-stage analysis procedure in accordance with ASCE/SEI 12.2.3.2.

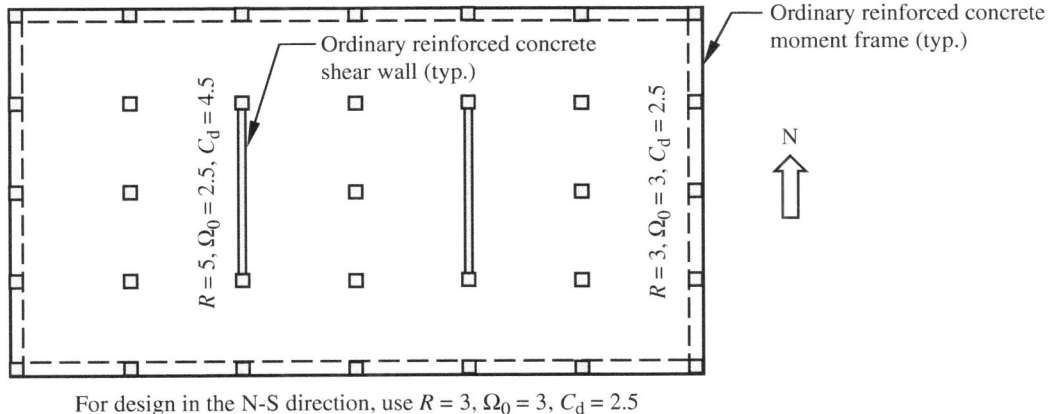

For design in the N-S direction, use $R = 3$, $\Omega_0 = 3$, $C_d = 2.5$

FIGURE 3.5 Horizontal combinations of SFRSs.

independent line of resistance. In other words, each SFRS in that direction is permitted to be designed for R of that system. The value of R that must be used in the design of the diaphragms must be the least R for any of the SFRSs in that direction.

3.3.4 Combination Framing Detailing Requirements

Structural members common to different SFRSs in any direction must be designed using the detailing requirements of ASCE/SEI Chapter 12 and the largest R of the connected SFRSs (ASCE/SEI 12.2.4). The intent is to preserve the integrity of all the SFRSs when subjected to an earthquake event.

3.3.5 System-Specific Requirements

Requirements for dual systems, cantilever column systems, inverted pendulum-type structures, and other systems are given in ASCE/SEI 12.2.5. These requirements must be satisfied in addition to the requirements in ASCE/SEI Table 12.2-1 and other requirements in ASCE/SEI Chapter 12.

For shear wall-frame interactive systems, the shear strength of the shear walls must be at least 75 percent of the design story shear at each story (ASCE/SEI 12.2.5.8). The frames must be capable of resisting at least 25 percent of the design story shear in each story. The frames are basically a backup system for the walls.

3.4 Diaphragm Flexibility, Configuration Irregularities, and Redundancy

3.4.1 Diaphragm Flexibility

The manner in which horizontal seismic forces get transferred to the vertical elements of the SFRS depends on the flexibility of the diaphragm. Descriptions of flexible, rigid, and semirigid diaphragms are given in Table 3.4.

Diaphragms are permitted to be idealized as flexible or rigid based on construction type (see ASCE/SEI 12.3.1.1, ASCE/SEI 12.3.1.2, and Table 3.5).

Flexible	In-plane deflection of the diaphragm due to horizontal forces is relatively large compared to the deflection of the vertical elements of the SFRS. The diaphragm does not undergo rigid body rotation when subjected to horizontal forces. Horizontal forces are distributed to the vertical elements of the SFRS in proportion to the mass of the diaphragm tributary to the vertical elements of the SFRS (for diaphragms of uniform material and weight, horizontal forces can be distributed in proportion to the area tributary to the vertical elements of the SFRS).
Rigid	In-plane deflection of the diaphragm due to horizontal forces is relatively small compared to the deflection of the vertical elements of the SFRS. The diaphragm displaces and rotates as a rigid body when subjected to horizontal forces. Horizontal forces are distributed to the vertical elements of the SFRS in proportion to the relative rigid ties (stiffnesses) and their location with respect to the center of rigidity (CR).
Semirigid	In-plane deflection of the diaphragm due to horizontal forces is of the same order of magnitude as the deflection of the vertical elements of the SFRS. Horizontal forces are distributed to the vertical elements of the SFRS based on a structural analysis that explicitly includes the in-plane stiffness of the diaphragm.

TABLE 3.4 Types of Diaphragms

Flexible	Untopped steel decking or wood structural panels provided any of the following conditions are satisfied: • The SFRS consists of (1) steel braced frames, (2) steel and concrete composite braced frames, or (3) concrete, masonry, steel, or steel and concrete composite shear walls. • The building is a one- or two-family dwelling. • Structures of light-frame construction where both of the following conditions are satisfied: (1) topping of concrete or similar materials is not placed over wood structural panel diaphragms except for nonstructural topping less than or equal to 1.5 in. (38 mm) thick and (2) each line of vertical elements of the SFRS complies with the allowable story drift requirements in ASCE/SEI Table 12.12-1.
Rigid	Concrete slabs or concrete-filled metal deck with (1) span-to-depth ratios of 3 or less (see Fig. 3.6) and (2) no structural irregularities as defined in ASCE/SEI Table 12.3-3.

TABLE 3.5 Idealized Diaphragm Flexibility

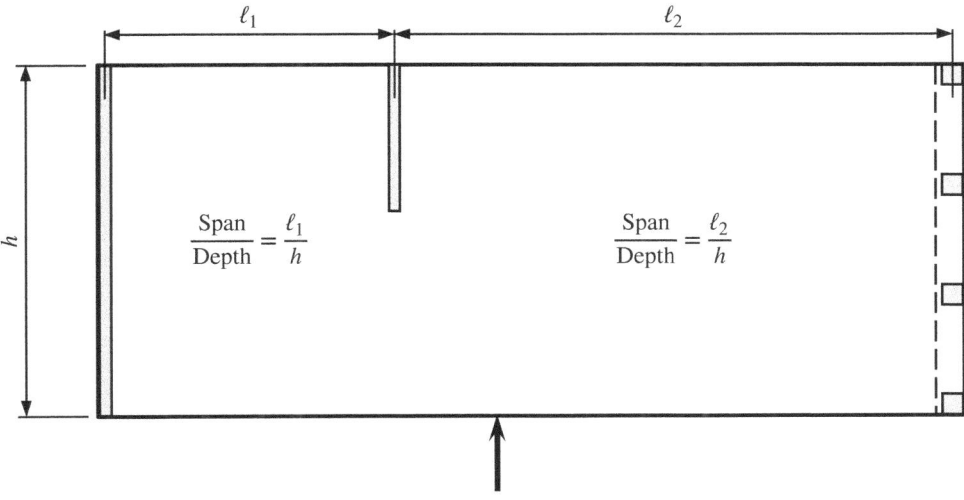

FIGURE 3.6 Definition of span-to-depth ratio for a diaphragm.

For diaphragms not satisfying the conditions in ASCE/SEI 12.3.1.1 for flexible diaphragms and ASCE/SEI 12.3.1.2 for rigid diaphragms, ASCE/SEI Equation (12.3-1) can be used to determine if a diaphragm can be idealized as flexible:

$$\frac{\delta_{MDD}}{\Delta_{ADVE}} > 2 \qquad (3.1)$$

In this equation, δ_{MDD} is the computed maximum in-plane deflection of the diaphragm subjected to horizontal forces and Δ_{ADVE} is the average drift of adjoining vertical elements of the SFRS over the story below the diaphragm under consideration subjected to a tributary horizontal force equivalent to that used in the computation of δ_{MDD} (see Fig. 3.7). Diaphragms satisfying this equation can be idealized as flexible.

3.4.2 Irregular and Regular Classifications

Overview
A regular structure can be defined as one that possesses a distribution of mass, stiffness, and strength that results in essentially a uniform sway when subjected to ground shaking. In other words, the lateral displacement in each story on each side of the structure will be about the same.

The dissipation of earthquake energy is basically uniform throughout a regular structure; this results in relatively light and well-distributed damage in the structure. In contrast, structures that are irregular can suffer extreme damage at only one or a few locations; this can result in localized failure of structural members and can lead to a loss of the structure's ability to survive the ground shaking. Past earthquakes have revealed that irregular structures suffer greater damage than those that are regular.

Structures are classified as regular or irregular based on the criteria in ASCE/SEI 12.3.2. Structural configurations can be divided into horizontal and vertical types

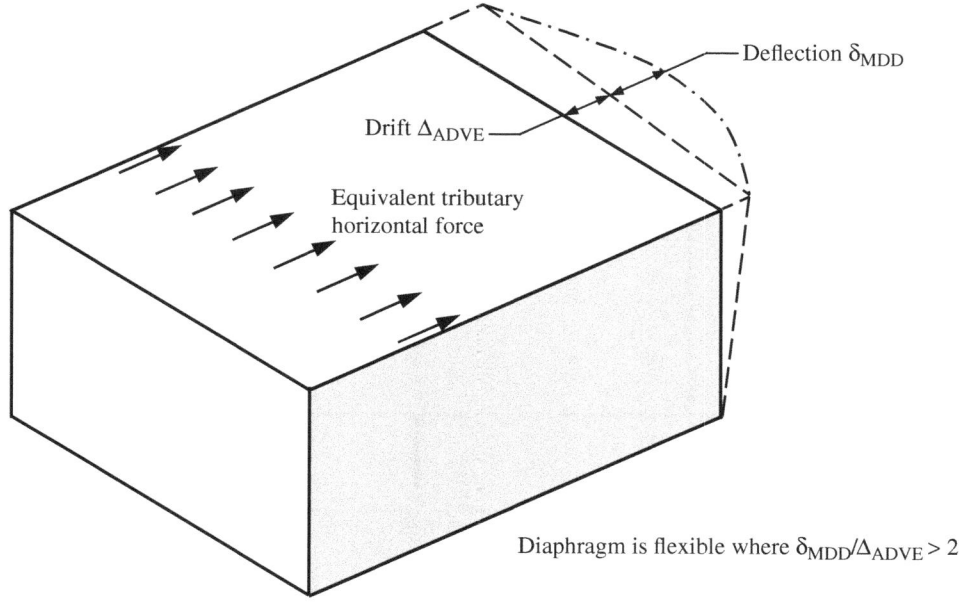

FIGURE 3.7   Definition of a flexible diaphragm in accordance with ASCE/SEI Equation (12.3-1).

for purposes of defining irregularities. Both types of irregularities are covered in the following text.

Most of the earthquake provisions in ASCE/SEI 7 are applicable to regular structures; for example, the ELF procedure in ASCE/SEI 12.8 using elastic analysis is not capable of accurately predicting the earthquake effects on certain types of irregular buildings. Limitations and additional requirements must be satisfied for irregular structures depending on the SDC.

Horizontal Irregularities

Structures with one of more of the irregularities defined in ASCE/SEI Table 12.3-1 are designated as having a horizontal structural irregularity (see Table 3.6).

Vertical Irregularities

Structures with one or more of the irregularities defined in ASCE/SEI Table 12.3-2 are designated as having a vertical structural irregularity (see Table 3.7).

3.4.3   Limitations and Additional Requirements for Systems with Structural Irregularities

The requirements in ASCE/SEI 12.3.3 are meant to account for possible detrimental effects caused by irregularities and to prohibit or limit certain types of irregularities in areas of high seismic risk.

A summary of these limitations and additional requirements is given in Table 3.8.

Type	Description		
1a	**Torsional Irregularity:** Applicable to rigid or semirigid diaphragms where the maximum story drift, Δ_{max}, computed including accidental torsion with $A_x = 1.0$, at one end of the structure transverse to an axis is more than 1.2 times the average of the story drifts at the two ends of the structure.	$\Delta_{avg} = \dfrac{\Delta_{max} + \Delta_{min}}{2}$ Δ_{max} Δ_{min} ↑ Seismic force Torsional irregularity: $\Delta_{max} > 1.2\Delta_{avg}$ Extreme torsional irregularity: $\Delta_{max} > 1.4\Delta_{avg}$	

ASCE/SEI Reference Section	SDC Application
12.3.3.4	D, E, and F
12.7.3	B, C, D, E, and F
12.8.4.3	C, D, E, and F
12.12.1	C, D, E, and F
Table 12.6-1	D, E, and F
16.3.4	B, C, D, E, and F

Type	Description
1b	**Extreme Torsional Irregularity:** Applicable to rigid or semirigid diaphragms where the maximum story drift, Δ_{max}, computed including accidental torsion with $A_x = 1.0$, at one end of the structure transverse to an axis is more than 1.4 times the average of the story drifts at the two ends of the structure.

ASCE/SEI Reference Section	SDC Application
12.3.3.1	E and F
12.3.3.4	D
12.3.4.2	D
12.7.3	B, C, and D
12.8.4.3	C and D
12.12.1	C and D
Table 12.6-1	D
16.3.4	B, C, and D

Type	Description
2	**Reentrant Corner Irregularity:** Both plan projections of the structure beyond a reentrant corner are greater than 15 percent of the plan dimension of the structure in the given direction.

Reentrant corner irregularity:
$L_{xp}/L_x > 0.15$ and $L_{yp}/L_y > 0.15$

ASCE/SEI Reference Section	SDC Application
12.3.3.4	D, E, and F
Table 12.6-1	D, E, and F

TABLE 3.6 Horizontal Structural Irregularities in Accordance with ASCE/SEI Table 12.3-1

Type	Description	
3	**Diaphragm Discontinuity Irregularity:** Diaphragm with an abrupt discontinuity or variation in stiffness, including diaphragms with a cutout or open area greater than 50 percent of the gross enclosed diaphragm area or a change in effective diaphragm stiffness of more than 50 percent from one story to the next. <table><tr><td>ASCE/SEI Reference Section</td><td>SDC Application</td></tr><tr><td>12.3.3.4</td><td>D, E, and F</td></tr><tr><td>Table 12.6-1</td><td>D, E, and F</td></tr></table>	Diaphragm discontinuity: $L_{x1}L_{y1} > 0.5\,L_xL_y$
4	**Out-of-Plane Offset Irregularity:** Discontinuity in a lateral force-resistance path, such as an out-of-plane offset of at least one of the vertical elements. <table><tr><td>ASCE/SEI Reference Section</td><td>SDC Application</td></tr><tr><td>12.3.3.3</td><td>B, C, D, E, and F</td></tr><tr><td>12.3.3.4</td><td>D, E, and F</td></tr><tr><td>12.7.3</td><td>B, C, D, E, and F</td></tr><tr><td>Table 12.6-1</td><td>D, E, and F</td></tr><tr><td>16.3.4</td><td>B, C, D, E, and F</td></tr></table>	
5	**Nonparallel System Irregularity:** Vertical lateral force–resisting elements are not parallel to the major orthogonal axes of the SFRS. <table><tr><td>ASCE/SEI Reference Section</td><td>SDC Application</td></tr><tr><td>12.5.3</td><td>C, D, E, and F</td></tr><tr><td>12.7.3</td><td>B, C, D, E, and F</td></tr><tr><td>Table 12.6-1</td><td>D, E, and F</td></tr><tr><td>16.3.4</td><td>B, C, D, E, and F</td></tr></table>	

TABLE 3.6 Horizontal Structural Irregularities in Accordance with ASCE/SEI Table 12.3-1 (*Continued*)

Type	Description
1a	**Stiffness—Soft Story Irregularity:** A story in which the lateral stiffness is less than 70 percent of that in the story above or less than 80 percent of the average stiffness of the 3 stories above.

ASCE/SEI Reference Section	SDC Application
Table 12.6-1	D, E, and F

Type	Description
1b	**Stiffness—Extreme Soft Story Irregularity:** A story in which the lateral stiffness is less than 60 percent of that in the story above or less than 70 percent of the average stiffness of the 3 stories above.

ASCE/SEI Reference Section	SDC Application
12.3.3.1	E and F
Table 12.6-1	D, E, and F

- Stiffness—Soft story irregularity:

$$K_1 < 0.7\, K_2 \text{ or } K_1 < 0.8\, (K_2 + K_3 + K_4)/3$$

- Stiffness—Extreme soft story irregularity:

$$K_1 < 0.6\, K_2 \text{ or } K_1 < 0.7\, (K_2 + K_3 + K_4)/3$$

Type	Description
2	**Weight Mass Irregularity:** Effective mass of any story is more than 150 percent of the effective mass of an adjacent story. A roof that is lighter than the floor below need not be considered.

ASCE/SEI Reference Section	SDC Application
Table 12.6-1	D, E, and F

Weight (mass) irregularity:

$$M_5 > 1.5\, M_4 \text{ or } M_5 > 1.5\, M_6$$

TABLE 3.7 Vertical Structural Irregularities in Accordance with ASCE/SEI Table 12.3-2

Type	Description	
3	**Vertical Geometric Irregularity:** Horizontal dimension of the SFRS in any story is more than 130 percent of that in an adjacent story.	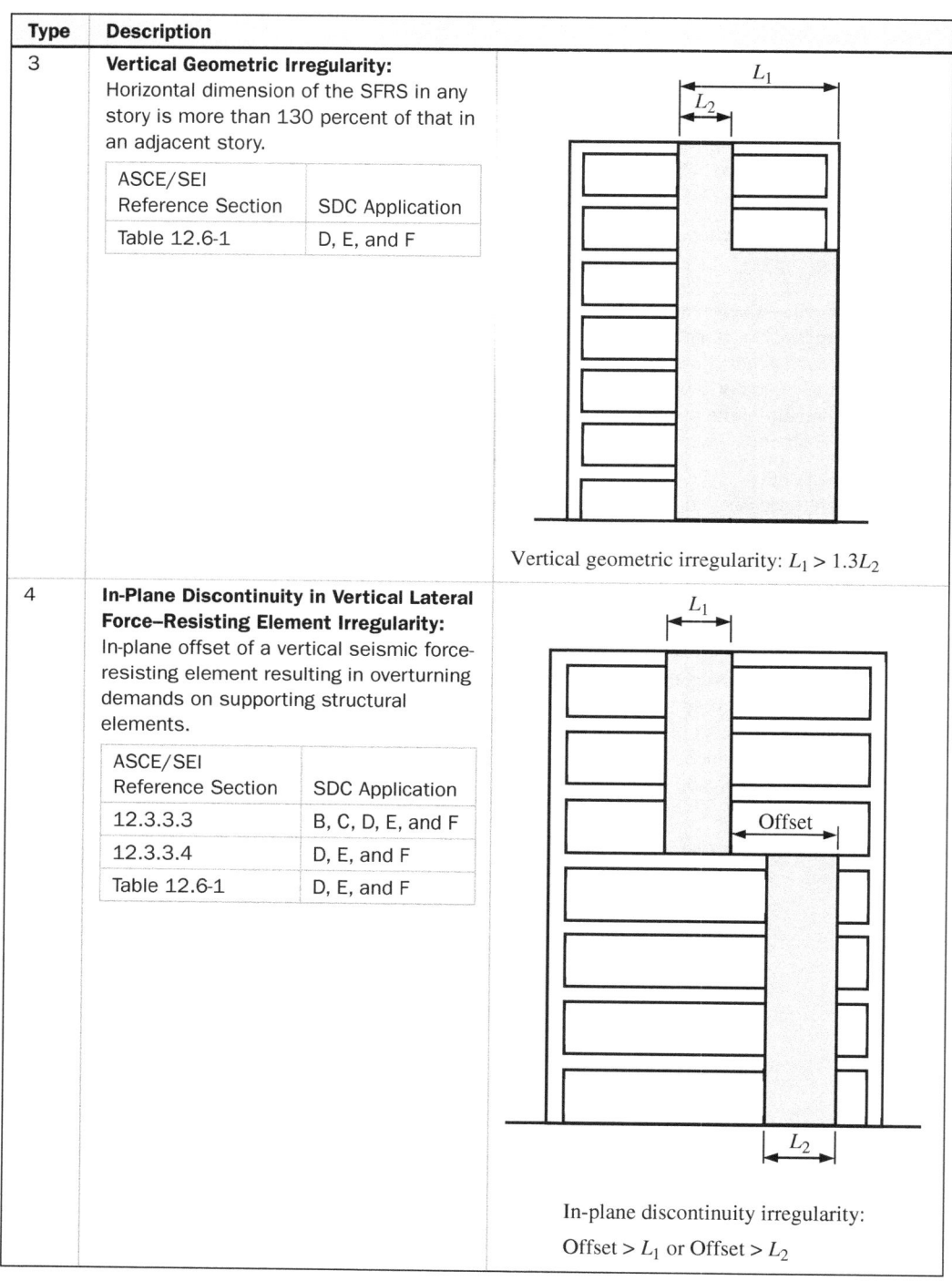

For Type 3:

ASCE/SEI Reference Section	SDC Application
Table 12.6-1	D, E, and F

For Type 4:

In-Plane Discontinuity in Vertical Lateral Force–Resisting Element Irregularity: In-plane offset of a vertical seismic force-resisting element resulting in overturning demands on supporting structural elements.

ASCE/SEI Reference Section	SDC Application
12.3.3.3	B, C, D, E, and F
12.3.3.4	D, E, and F
Table 12.6-1	D, E, and F

Vertical geometric irregularity: $L_1 > 1.3L_2$

In-plane discontinuity irregularity: Offset $> L_1$ or Offset $> L_2$

TABLE 3.7 Vertical Structural Irregularities in Accordance with ASCE/SEI Table 12.3-2 (*Continued*)

Type	Description			
5a	**Discontinuity in Lateral Strength—Weak Story Irregularity:** Story lateral strength is less than 80 percent of that in the story above. The story lateral strength is the total lateral strength of all seismic-resisting elements sharing the story shear for the direction under consideration.			

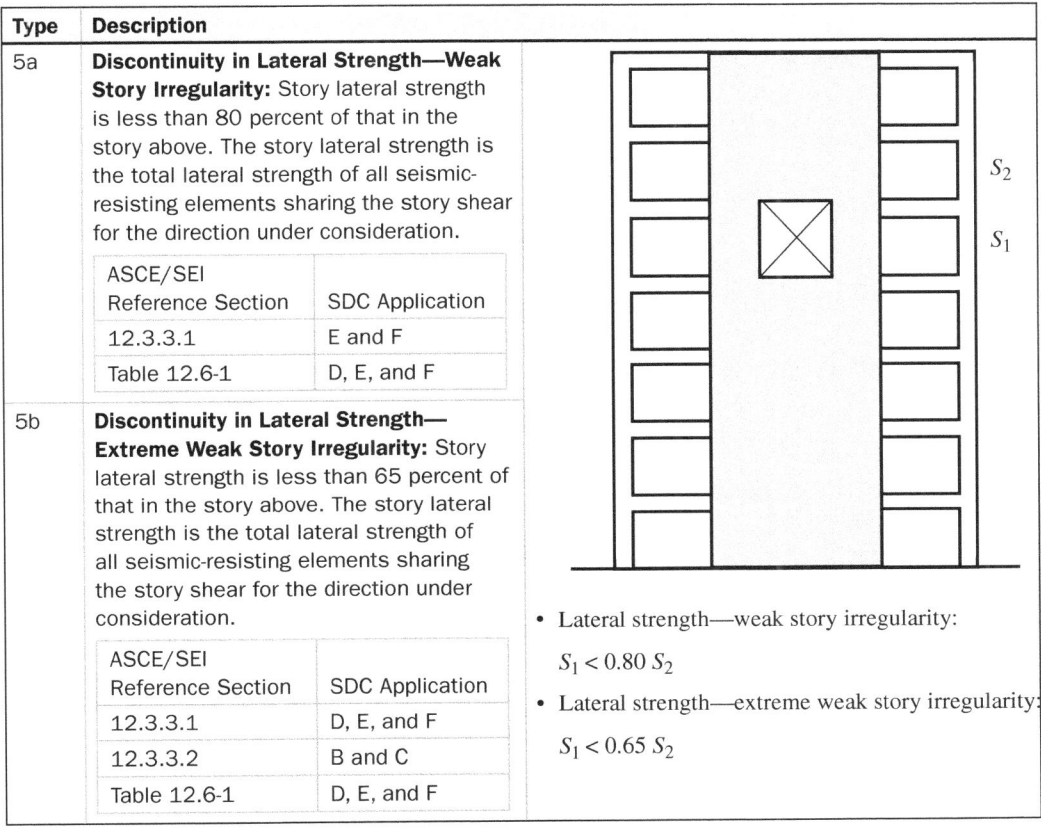

	ASCE/SEI Reference Section	SDC Application
	12.3.3.1	E and F
	Table 12.6-1	D, E, and F

Type	Description
5b	**Discontinuity in Lateral Strength—Extreme Weak Story Irregularity:** Story lateral strength is less than 65 percent of that in the story above. The story lateral strength is the total lateral strength of all seismic-resisting elements sharing the story shear for the direction under consideration.

	ASCE/SEI Reference Section	SDC Application
	12.3.3.1	D, E, and F
	12.3.3.2	B and C
	Table 12.6-1	D, E, and F

- Lateral strength—weak story irregularity:

$$S_1 < 0.80\, S_2$$

- Lateral strength—extreme weak story irregularity:

$$S_1 < 0.65\, S_2$$

TABLE 3.7 Vertical Structural Irregularities in Accordance with ASCE/SEI Table 12.3-2 (*Continued*)

3.4.4 Redundancy

The redundancy factor, ρ, is a measure of the redundancy inherent in a structure. A higher degree of redundancy exists where there are multiple paths to resist the lateral forces. A redundancy factor equal to 1.0 means a structure is sufficiently redundant, and no increase in the seismic load effects is warranted. For less redundant structures, ρ is equal to 1.3 (ASCE/SEI 12.3.4); this essentially reduces R, which, in turn, increases the seismic demand on the structural system.

The value of ρ is permitted to equal 1.0 in the following cases:

1. Structures assigned to SDC B or C.
2. Drift calculations and P-delta effects.
3. Design of nonstructural components.
4. Design of nonbuilding structures not similar to buildings.
5. Design of collector elements, splices, and their connections for which the seismic load effects, including overstrength of ASCE/SEI 12.4.3, are used.

	SDC	Irregularity Type	Limitations/Additional Requirements
Prohibited horizontal and vertical irregularities for SDCs D, E, or F	D	Vertical irregularity type 5b	Not permitted
	E and F	• Horizontal irregularity type 1b • Vertical irregularities type 1b, 5a, or 5b	
Extreme weak stories	B and C	Vertical irregularity type 5b	Height is limited to 2 stories or 30 ft (9.1 m)[1]
Elements supporting discontinuous walls or frames	B through F	• Horizontal irregularity type 4 • Vertical irregularity type 4	Supporting members and their connections must be designed to resist load combinations with overstrength factor of ASCE/SEI 12.4.3
Increase in forces due to irregularities for SDCs D, E, and F	D through F	• Horizontal irregularities type 1a, 1b, 2, 3, or 4 • Vertical irregularity type 4	Design forces determined by ASCE/SEI 12.10.1.1 must be increased by 1.25 for the following: (a) Connections of diaphragms to vertical elements and to collectors (b) Collectors and their connections, including connections of collectors to vertical elements of the SFRS[2]
Incorporation of accidental torsional moment defined in ASCE/SEI 12.8.4.2 in the analysis	B through F	Horizontal irregularities in ASCE/SEI Table 12.3-1	Accidental torsional moment, M_{ta}, need not be included when determining (1) seismic forces, E, in the design of the structure and (2) design story drifts in ASCE/SEI 12.8.6, 12.9.1.2, or ASCE/SEI Chapter 16, or drift limits of ASCE/SEI 12.12.1 except for the following structures: (a) Structures assigned to SDC B with type 1b horizontal irregularity (b) Structures assigned to SDC C, D, E, and F with type 1a or 1b horizontal irregularity

(1) This limit does not apply where the weak story is capable of resisting a total seismic force equal to Ω_0 times the design force determined in accordance with ASCE/SEI 12.8.

(2) Forces calculated using the seismic load effects, including the overstrength factor of ASCE/SEI 12.4.3, need not be increased.

TABLE 3.8   Limitations and Additional Requirements for Systems with Structural Irregularities

6. Design of members or connections where the seismic load effects, including overstrength of ASCE/SEI 12.4.3, are required for design.

7. Diaphragm loads determined by ASCE/SEI Equation (12.10-1), including the limits imposed by ASCE/SEI Equations (12.10-2) and (12.10-3).

8. Structures with damping systems designed in accordance with ASCE/SEI Chapter 18.

9. Design of walls for out-of-plane forces, including their anchorage.

For structures assigned to SDC D through F that do not satisfy any of the conditions listed above, $\rho = 1.3$ unless one of the two conditions in ASCE/SEI 12.3.4.2 is met in which case $\rho = 1.0$.

In the first condition, the applicable requirements in ASCE/SEI Table 12.3-3 must be satisfied in each story in a structure that resists more than 35 percent of the base shear in the direction of analysis. The intent is to exclude the uppermost stories and penthouses in a structure from the redundancy requirements.

The requirements under the first condition are organized with respect to the following lateral force–resisting elements:

- Braced frames
- Moment frames
- Shear walls or wall piers with a height-to-length ratio greater than 1.0 (see ASCE/SEI Figure 12.3-2)
- Cantilever columns
- Other

A dual system is included under the "Other" types of elements where no requirements need to be satisfied (that is, $\rho = 1.0$) because this system is considered to be inherently redundant.

In this approach, individual elements of the SFRS are removed to determine the effects on the remaining structure. If the removal of such elements, one by one, does not result in (1) more than a 33 percent reduction in story strength (or, in the case of cantilever columns, a 33 percent reduction in moment resistance) or (2) an extreme torsional irregularity (horizontal structural irregularity type 1b), ρ may be taken as 1.0. This check determines whether an individual member plays a significant role in lateral force resistance.

In the second condition, ρ is permitted to be taken as 1.0 for structures regular in plan at all levels provided the SFRSs consist of at least two bays of seismic force–resisting perimeter framing on each side of the structure in each orthogonal direction at each story resisting more than 35 percent of the base shear. In the case of shear walls, the number of bays is taken equal to the length of the shear wall divided by the story height in all cases except where light-frame construction is used; in that case, the number of bays is equal to two times the length of the shear wall divided by the story height (see ASCE/SEI Figure 12.3-2).

It is permitted to use $\rho = 1.3$ without checking the two conditions in ASCE/SEI 12.3.4.2; using this default value may not result in the most economical solution.

The flowchart in Fig. 3.8 can be used to determine ρ.

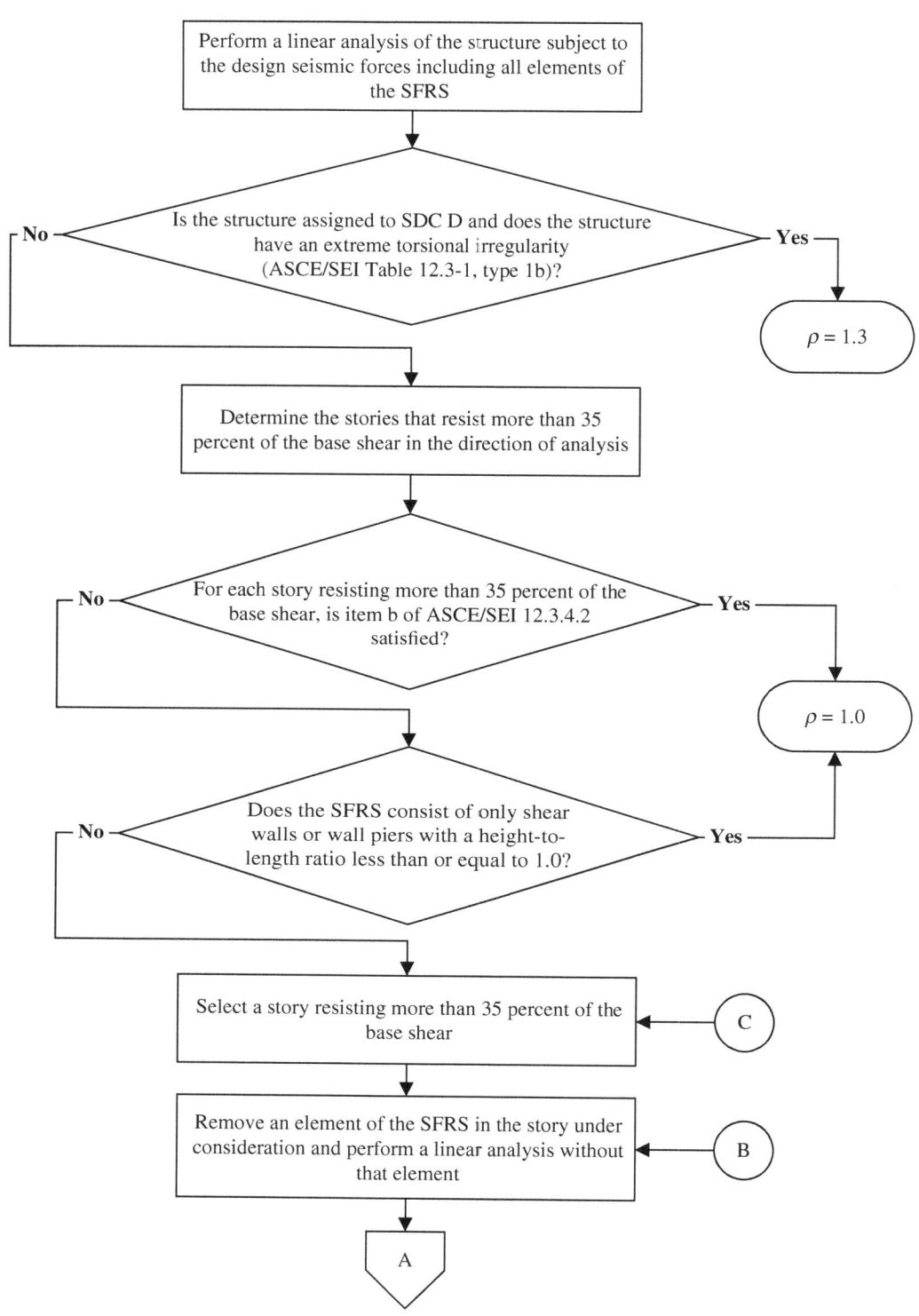

FIGURE 3.8 Calculation of ρ for structures assigned to SDC D through F.

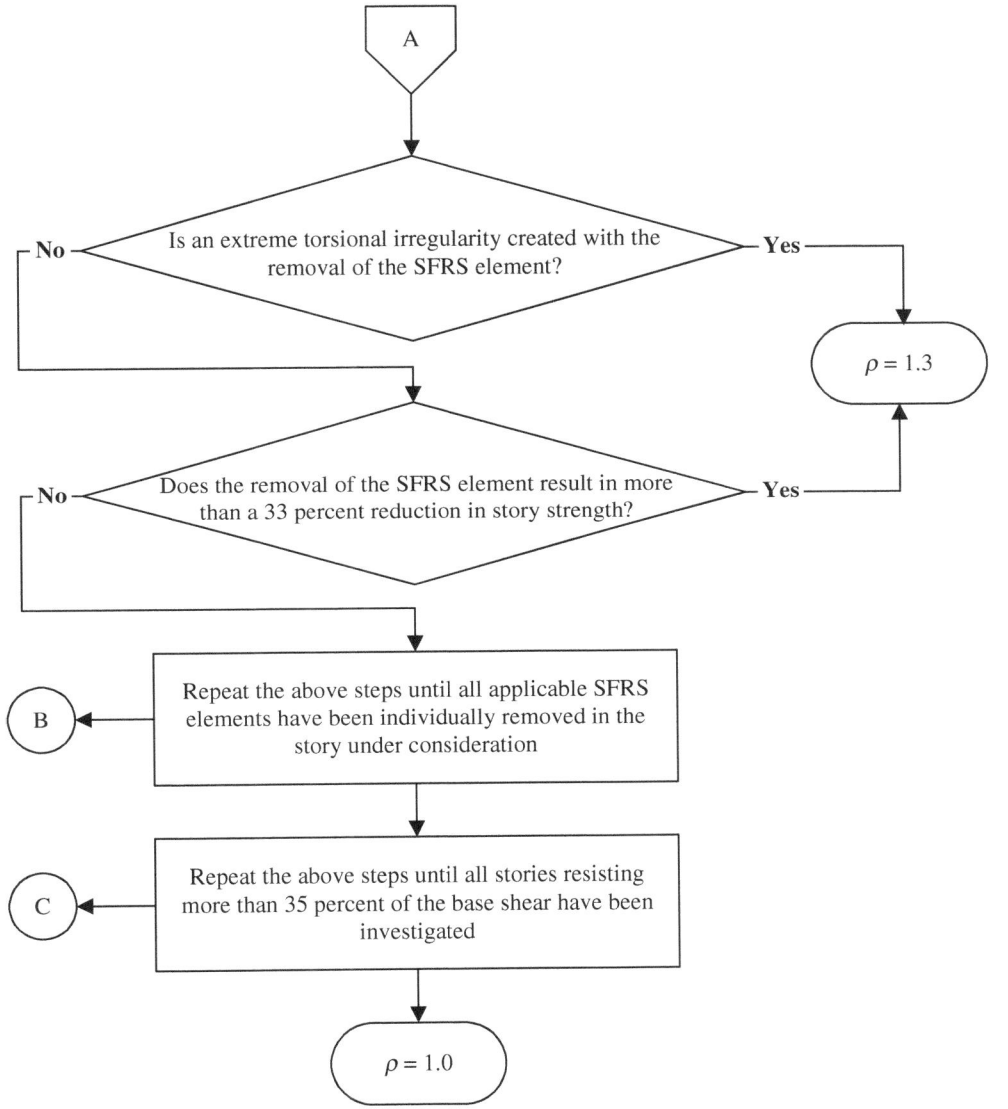

FIGURE 3.8   *(Continued)*

3.5   Seismic Load Effects and Combinations

3.5.1   Overview

Unless otherwise exempted, all structural members, including members that are not part of the SFRS, must be designed using the seismic load effects of ASCE/SEI 12.4. These effects include axial forces, shear forces, and bending moments due to the application of horizontal and vertical forces on the structure (ASCE/SEI 12.4.2).

Certain members must be designed for seismic load effects including the over-strength factor, Ω_0 (ASCE/SEI 12.4.3). The conditions under which these effects are applicable are given below.

3.5.2   Seismic Load Effect, *E*

The seismic load effect, *E*, used in the load combinations in ASCE/SEI 2.3 and 2.4 is the sum of effects of horizontal (E_h) and vertical (E_v) seismic forces, which are determined by ASCE/SEI Equations (12.4-3) and (12.4-4), respectively:

$$E_h = \rho Q_E \tag{3.2}$$

$$E_v = 0.2S_{DS}D \tag{3.3}$$

In Eq. (3.2), Q_E are the effects (axial forces, shear forces, and bending moments) obtained from the structural analysis from either the base shear, *V*, which is distributed over the height of the structure, or the seismic force acting on a component of a structure, F_p (ASCE/SEI 12.4). The effect from the vertical seismic forces, E_v, is based on an assumed effective vertical acceleration of $0.2S_{DS}$ times the acceleration due to gravity. According to the first exception in ASCE/SEI 12.4.2.2, E_v is to be determined by ASCE/SEI Equation (12.4-4b) where the option to incorporate the effects of vertical ground motions by ASCE/SEI 11.9 is required:

$$E_v = 0.3S_{av}D \tag{3.4}$$

In this equation, S_{av} is the design vertical response spectral acceleration determined by ASCE/SEI 11.9.3 (see Sec. 2.7 of this publication).

The second exception in ASCE/SEI 12.4.2.2 permits E_v to be taken as zero for either of the following cases:

1. In ASCE/SEI Equations (12.4-1), (12.4-2), (12.4-5), and (12.4-6) for structures assigned to SDC B.

2. In ASCE/SEI Equation (12.4-2) when determining demands on the soil-structure interface of foundations.

It is important to understand the proper sign convention to use in the load combinations with *E*. In the strength design load combination 6 in ASCE/SEI 2.3.6 or the allowable stress load combinations 8 and 9 in ASCE/SEI 2.4.5, *E* is determined by ASCE/SEI Equation (12.4-1):

$$E = E_h + E_v = \rho Q_E + 0.2S_{DS}D \tag{3.5}$$

In these load combinations, effects from the dead load, *D*, are additive to those due to *E*.

In the strength design load combination 7 in ASCE/SEI 2.3.6 or the allowable stress load combination 10 in ASCE/SEI 2.4.5, *E* is determined by ASCE/SEI (12.4-2):

$$E = E_h - E_v = \rho Q_E - 0.2S_{DS}D \tag{3.6}$$

In these load combinations, the effects of *E* counteract those from *D*.

Basic load combinations for strength design and allowable stress design based on SDC are given in Table 3.9.

ASCE/SEI 7 Load Combination		Load Combination
Strength Design*		
6	SDC B	$1.2D + Q_E + L + 0.2S$
7		$0.9D + Q_E$
6	SDC C	$(1.2 + 0.2S_{DS})D + Q_E + L + 0.2S$
7		$(0.9 - 0.2S_{DS})D + Q_E$
6	SDC D, E, and F	$(1.2 + 0.2S_{DS})D + \rho Q_E + L + 0.2S$
7		$(0.9 - 0.2S_{DS})D + \rho Q_E$
Allowable Stress Design**		
8	SDC B	$D + 0.7Q_E$
9		$D + 0.525Q_E + 0.75L + 0.75S$
10		$0.6D + 0.7Q_E$
8	SDC C	$(1.0 + 0.14S_{DS})D + 0.7Q_E$
9		$(1.0 + 0.105S_{DS})D + 0.525Q_E + 0.75L + 0.75S$
10		$(0.6 - 0.14S_{DS})D + 0.7Q_E$
8	SDC D, E, and F	$(1.0 + 0.14S_{DS})D + 0.7\rho Q_E$
9		$(1.0 + 0.105S_{DS})D + 0.525\rho Q_E + 0.75L + 0.75S$
10		$(0.6 - 0.14S_{DS})D + 0.7\rho Q_E$

*Notes for strength design load combinations:
(1) Load factor on L in combination 6 is permitted to equal 0.5 for all occupancies in which L_o in ASCE/SEI Table 4.3-1 is less than or equal to 100 lb/ft² (4.79 kN/m²), with the exception of garages or areas occupied as places of public assembly.
(2) Where fluid loads, F, are present, they shall be included with the same load factor as dead load, D, in combinations 6 and 7.
(3) Where load H is present, it must be included as follows:
   (a) Where the effect of H adds to the primary variable load effect, include H with a load factor of 1.6.
   (b) Where the effect of H resists the primary variable load effect, include H with a load factor of 0.9 where the load is permanent or a load factor of 0 for all other conditions.
**Notes for allowable stress design load combinations:
(1) Where fluid loads, F, are present, they shall be included in combinations 8, 9, and 10 with the same factor as that used for dead load, D.
(2) Where load H is present, it must be included as follows:
   (a) Where the effect of H adds to the primary variable load effect, include H with a load factor of 1.0.
   (b) Where the effect of H resists the primary variable load effect, include H with a load factor of 0.6 where the load is permanent or a load factor of 0 for all other conditions.

TABLE 3.9 Seismic Load Combinations

3.5.3 Seismic Load Effects Including Overstrength

The seismic load effect including overstrength, E_m, used in the load combinations in ASCE/SEI 2.3 and 2.4 consists of effects of horizontal seismic forces including overstrength (E_{mh}) and vertical seismic load effect (E_v), the latter of which is determined by

Eq. (3.3). The horizontal seismic load effect with overstrength is determined by ASCE/SEI Equation (12.4-7):

$$E_{mh} = \Omega_0 Q_E \tag{3.7}$$

where Ω_0 is the overstrength factor given in ASCE/SEI Table 12.2-1 as a function of the SFRS (ASCE/SEI 12.4.3).

In the strength design load combination 6 in ASCE/SEI 2.3.6 or the allowable stress load combinations 8 and 9 in ASCE/SEI 2.4.5, E must be taken equal to E_m determined by ASCE/SEI Equation (12.4-5):

$$E_m = E_{mh} + E_v = \Omega_0 Q_E + 0.2 S_{DS} D \tag{3.8}$$

In the strength design load combination 7 in ASCE/SEI 2.3.6 or the allowable stress load combination 10 in ASCE/SEI 2.4.5, E must be taken equal to E_m determined by ASCE/SEI Equation (12.4-6):

$$E_m = E_{mh} - E_v = \Omega_0 Q_E - 0.2 S_{DS} D \tag{3.9}$$

Where capacity-limited design is required, the capacity-limited horizontal seismic load, E_{cl}, which is the maximum force that can develop in the element determined by a rational, plastic mechanism analysis, must be substituted for E_{mh} in the above equations (ASCE/SEI 12.4.3.2).

Basic combinations for strength design with overstrength and allowable stress design with overstrength based on SDC are given in Table 3.10. These combinations pertain only to the following structural elements:

- Elements supporting discontinuous walls or frames in structures assigned to SDC B through F.
- Collector elements and their connections, including connections to vertical elements in structures assigned to SDC C through F.

Allowable stresses are permitted to be determined using an allowable stress increase factor of 1.2 where seismic loads including overstrength in ASCE/SEI 12.4.3 are used in load combinations 8, 9, and 10; this increase is not to be combined with any other increases in allowable stresses or load combination reductions permitted in any of the provisions (ASCE/SEI 2.4.5).

3.5.4 Minimum Upward Forces for Horizontal Cantilevers (SDCs D through F)

Horizontal cantilever members in buildings assigned to SDC D, E, or F must be designed for a minimum net upward force of $0.2D$ in addition to the applicable load combinations in ASCE/SEI 12.4. This requirement is meant to provide minimum strength in the upward direction and to account for any dynamic amplification of vertical ground motion due to the vertical flexibility of the cantilever.

3.6 Direction of Loading

Earthquakes can produce ground motion in any direction and a structure must be designed to resist the maximum possible effects. Seismic forces must be applied to the structure in directions producing the most critical load effects on the structural members (ASCE/SEI 12.5). Direction of loading requirements are given in Table 3.11.

ASCE/SEI 7 Load Combination		Load Combination
Strength Design*		
6	SDC B	$1.2D + \Omega_0 Q_E + L + 0.2S$
7		$0.9D + \Omega_0 Q_E$
6	SDC C, D, E, and F	$(1.2 + 0.2S_{DS})D + \Omega_0 Q_E + L + 0.2S$
7		$(0.9 - 0.2S_{DS})D + \Omega_0 Q_E$
Allowable Stress Design**		
8	SDC B	$D + 0.7\Omega_0 Q_E$
9		$D + 0.525\Omega_0 Q_E + 0.75L + 0.75S$
10		$0.6D + 0.7\Omega_0 Q_E$
8	SDC C, D, E, and F	$(1.0 + 0.14S_{DS})D + 0.7\Omega_0 Q_E$
9		$(1.0 + 0.105S_{DS})D + 0.525\Omega_0 Q_E + 0.75L + 0.75S$
10		$(0.6 - 0.14S_{DS})D + 0.7\Omega_0 Q_E$

*Notes for strength design load combinations:
(1) Load factor on L in combination 6 is permitted to equal 0.5 for all occupancies in which L_o in ASCE/SEI Table 4.3-1 is less than or equal to 100 lb/ft² (4.79 kN/m²), with the exception of garages or areas occupied as places of public assembly.
(2) Where fluid loads, F, are present, they shall be included with the same load factor as dead load, D, in combinations 6 and 7.
(3) Where load H is present, it must be included as follows:
   (a) Where the effect of H adds to the primary variable load effect, include H with a load factor of 1.6.
   (b) Where the effect of H resists the primary variable load effect, include H with a load factor of 0.9 where the load is permanent or a load factor of 0 for all other conditions.
**Notes for allowable stress design load combinations:
(1) Where fluid loads, F, are present, they shall be included in combinations 8, 9, and 10 with the same factor as that used for dead load, D.
(2) Where load H is present, it must be included as follows:
   (a) Where the effect of H adds to the primary variable load effect, include H with a load factor of 1.0.
   (b) Where the effect of H resists the primary variable load effect, include H with a load factor of 0.6 where the load is permanent or a load factor of 0 for all other conditions.

TABLE 3.10 Seismic Load Combinations with Overstrength

Orthogonal effects on horizontal members such as beams and slabs are generally minimal while those on vertical elements (columns and walls) that are part of the SFRS can be substantial to the point of governing the design. The maximum effects Q_E obtained by the methods in Table 3.11 must be modified by ρ or Ω_0, whichever is applicable, and then combined with E_v accordingly.

3.7 Analysis Procedure Selection

Requirements on the type of procedure permitted to be used to analyze structures for seismic load effects are given in ASCE/SEI 12.6 (see Table 3.12 and Fig. 3.9). Permitted analytical procedures depend on SDC, risk category, height and period of the structure, and the presence of any structural irregularities.

SDC	Requirement
B	Design seismic forces are permitted to be applied independently in each of two orthogonal directions and orthogonal interaction effects are permitted to be neglected.
C	(1) Conform to the requirements of SDC B. (2) Structures with horizontal irregularity type 5 (nonparallel system irregularity) must be analyzed using one of the following procedures: • Orthogonal combination procedure: Apply 100 percent of the seismic forces in one direction and 30 percent of the seismic forces in the perpendicular direction on the structure simultaneously where the forces are computed in accordance with ASCE/SEI 12.8 (equivalent lateral force procedure), ASCE/SEI 12.9.1 (modal response spectrum analysis procedure) or ASCE/SEI 12.9.2 (linear response history procedure) • Simultaneous application of orthogonal ground motion: Apply orthogonal pairs of ground motion acceleration histories simultaneously to the structure using ASCE/SEI 12.9.2 (linear response history procedure) or ASCE/SEI 12.6 (nonlinear response history procedure).
D–F	(1) Conform to the requirements of SDC C. (2) Any column or wall that forms part of two or more intersecting SFRSs and is subjected to axial load due to seismic forces along either principal axis greater than or equal to 20 percent of the axial design strength of the column or wall must be designed for the most critical load effect due to application of seismic forces in any direction. Either of the procedures of ASCE/SEI 12.5.3.a or 12.5.3.b is permitted to be used to satisfy this requirement. Except as required by ASCE/SEI 12.7.3, a two-dimensional analysis is permitted for structures with flexible diaphragms.

TABLE 3.11   Direction of Loading Requirements

3.8   Modeling Criteria

Requirements pertaining to the construction of an adequate model for the purposes of seismic load analysis are given in ASCE/SEI 12.7. Included are provisions for foundation modeling, effective seismic weight, structural modeling, and interaction effects. A summary of these requirements are given in Table 3.13.

3.9   Equivalent Lateral Force Procedure

3.9.1   Overview

The equivalent lateral force (ELF) procedure is a simplified method of determining the effects of ground motion on a structure. It is valid for essentially regular structures without certain types of significant discontinuities where the primary response to ground motion is in the horizontal direction without substantial torsion (that is, the first mode of vibration).

The effects of inelastic dynamic response are determined by using a linear static analysis: a design base shear V is determined, which is distributed over the height of

SDC	Structural Characteristics	Equivalent Lateral Force Procedure (ASCE/SEI 12.8)	Analysis Methods	
			Modal Response Spectrum Analysis (ASCE/SEI 12.9.1) or Linear Response History Analysis (ASCE/SEI 12.9.2)	Nonlinear Response History Procedures (ASCE/SEI Chapter 16)
B, C	All structures	P	P	P
D, E, F	Risk category I or II buildings not exceeding 2 stories above the base	P	P	P
	Structures of light-frame construction	P	P	P
	Structures with no structural irregularities and not exceeding 160 ft (48.8 m) in structural height	P	P	P
	Structures exceeding 160 ft (48.8 m) in structural height with no structural irregularities and with $T < 3.5T_s$	P	P	P
	Structures not exceeding 160 ft (48.8 m) in structural height and having only horizontal irregularities of type 2, 3, 4, or 5 in ASCE/SEI Table 12.3-1 or vertical irregularities of type 4, 5a, or 5b in ASCE/SEI Table 12.3-2	P	P	P
	All other structures	NP	P	P

P = permitted, NP = not permitted.

TABLE 3.12   Permitted Analytical Procedures

the structure at each floor level. The structure is analyzed for these static forces, which are distributed to the members of the SFRS considering the flexibility of the diaphragms.

The requirements of the ELF procedure are given in ASCE/SEI 12.8. This analysis method can be used for all structures assigned to SDC B and C as well as some types of structures assigned to SDC D, E, and F (see Table 3.12).

3.9.2   Seismic Base Shear

The seismic base shear, V, is determined by ASCE/SEI Equation (12.8-1):

$$V = C_s W \tag{3.10}$$

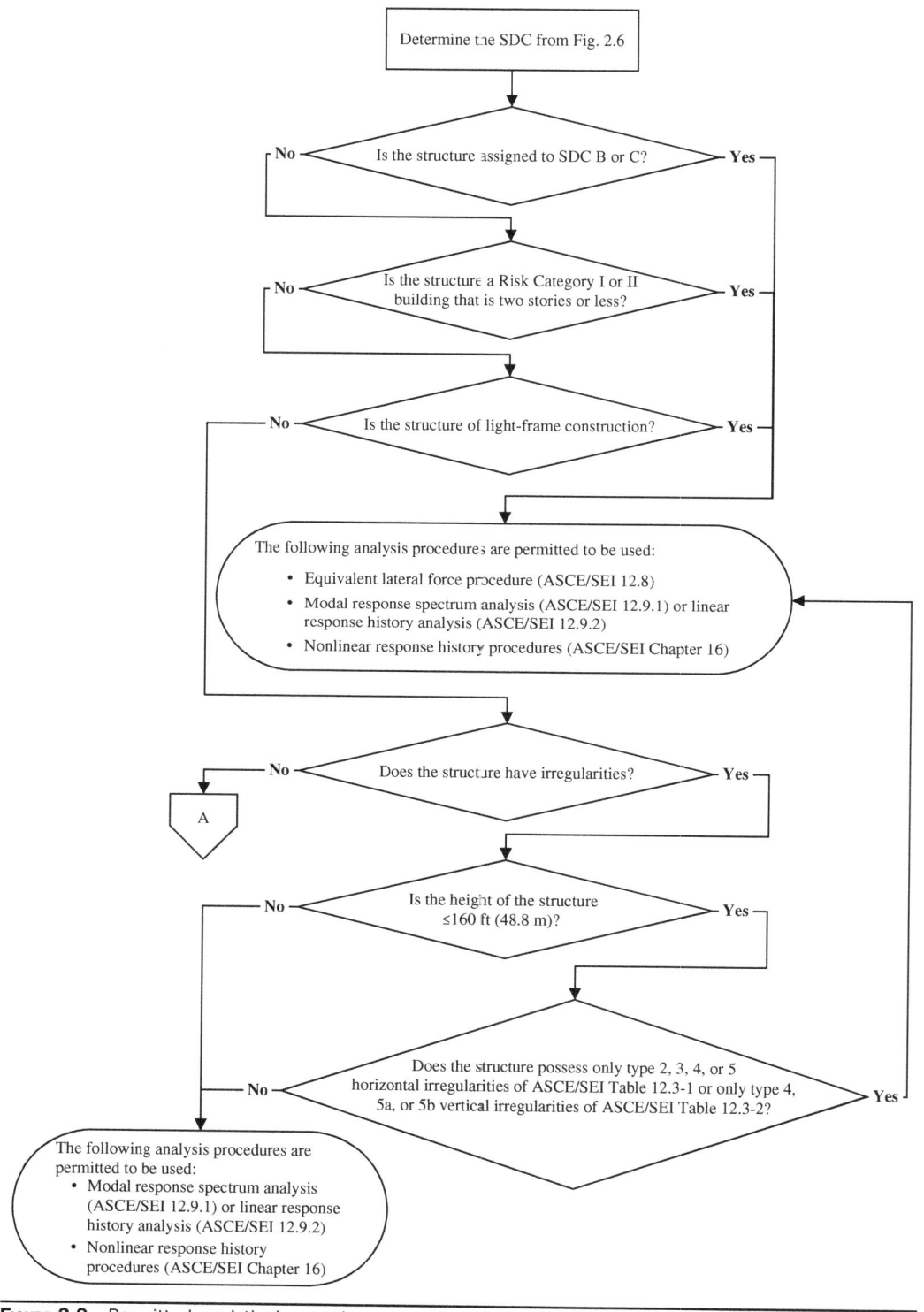

Determine the SDC from Fig. 2.6

No — Is the structure assigned to SDC B or C? — Yes

No — Is the structure a Risk Category I or II building that is two stories or less? — Yes

No — Is the structure of light-frame construction? — Yes

The following analysis procedures are permitted to be used:

- Equivalent lateral force procedure (ASCE/SEI 12.8)
- Modal response spectrum analysis (ASCE/SEI 12.9.1) or linear response history analysis (ASCE/SEI 12.9.2)
- Nonlinear response history procedures (ASCE/SEI Chapter 16)

No — Does the structure have irregularities? — Yes

A

No — Is the height of the structure ≤160 ft (48.8 m)? — Yes

No — Does the structure possess only type 2, 3, 4, or 5 horizontal irregularities of ASCE/SEI Table 12.3-1 or only type 4, 5a, or 5b vertical irregularities of ASCE/SEI Table 12.3-2? — Yes

The following analysis procedures are permitted to be used:
- Modal response spectrum analysis (ASCE/SEI 12.9.1) or linear response history analysis (ASCE/SEI 12.9.2)
- Nonlinear response history procedures (ASCE/SEI Chapter 16)

FIGURE 3.9 Permitted analytical procedures in accordance with ASCE/SEI 12.6.

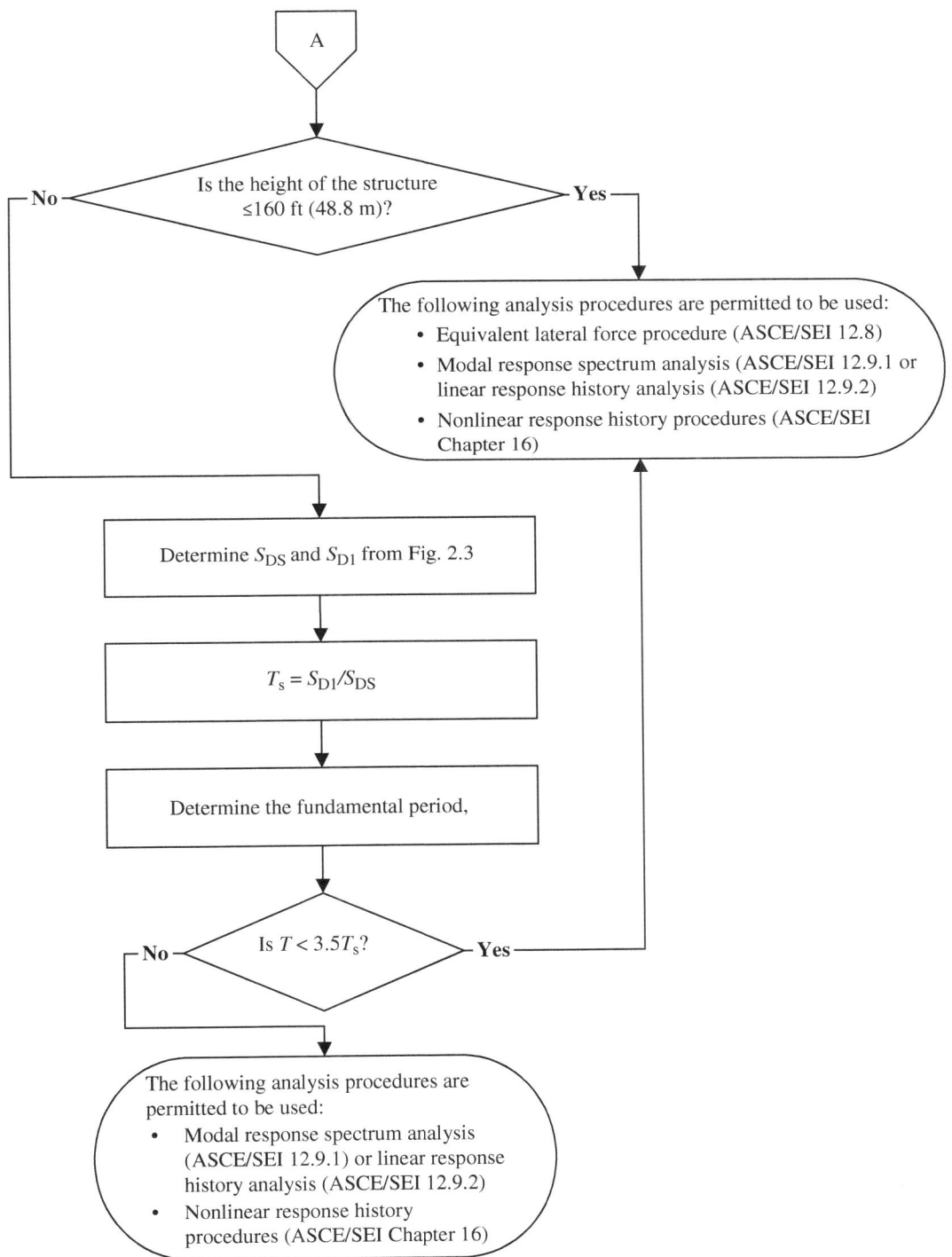

FIGURE 3.9   *(Continued)*

Item	ASCE/SEI Section Number	Requirements
Foundation modeling	12.7.1	• It is permitted to assume that the base of the structure is fixed for purposes of determining the seismic loads. • If the flexibility of the foundation is considered, the requirements of ASCE/SEI 12.13.3 (foundation load-deformation characteristics) or ASCE/SEI Chapter 19 (soil-structure interaction for seismic design) must be satisfied. • The base of a structure is defined as the level at which the horizontal seismic ground motions are considered to be imparted to the structure (ASCE/SEI 11.2). Additional information on where the base occurs for a number of common situations is given in ASCE/SEI C11.2.
Effective seismic weight, W	12.7.2	The following loads must be included in W: 1. In areas used for storage, a minimum of 25 percent of the floor live load, except for the following: (a) where the inclusion of storage loads adds no more than 5 percent to the effective seismic weight at that level, it need not be included in W and (b) floor live load in public garages and open parking structures need not be included. 2. Where partitions must be included in accordance with ASCE/SEI 4.3.2, the actual partition weight or 10 lb/ft² (0.48 kN/m²) of floor area, whichever is greater. 3. Total operating weight of permanent equipment. 4. Where the flat roof snow load, p_f, exceeds 30 lb/ft² (1.44 kN/m²), 20 percent of the uniform design snow load, regardless of the actual roof slope. 5. Weight of landscaping and other materials at roof gardens and similar areas.
Structural modeling	12.7.3	• A mathematical model of the structure including member stiffness and strength must be constructed for purposes of determining member forces and structure displacements due to applied loads and any imposed displacements or P-delta effects. • Cracked section properties must be used when analyzing concrete and masonry structures. In steel moment frames, the contribution of panel zone deformations to overall story drift must be included. • A three-dimensional analysis is required for structures that have horizontal irregularity types 1a, 1b, 4, or 5. A minimum of three dynamic degrees of freedom (translation in two orthogonal directions and torsional rotation about the vertical axis) must be included at each level of the structure. Structures with flexible diaphragms and type 4 horizontal structural irregularities need not be analyzed using a three-dimensional model. • The model must include representation of diaphragm stiffness characteristics where the diaphragm is not classified as rigid or flexible in accordance with ASCE/SEI 12.3.1. • When a modal response spectrum or response history analysis is performed, a minimum of three dynamic degrees of freedom (translation in two orthogonal plan directions and torsional rotation about the vertical axis) at each level must be used.
Interaction effects	12.7.4	Interaction effects between moment-resisting frames that are part the SFRS and rigid elements that are not part of the SFRS that enclose or adjoin the moment-resisting frame must be considered in the analysis.

TABLE 3.13 Modeling Criteria

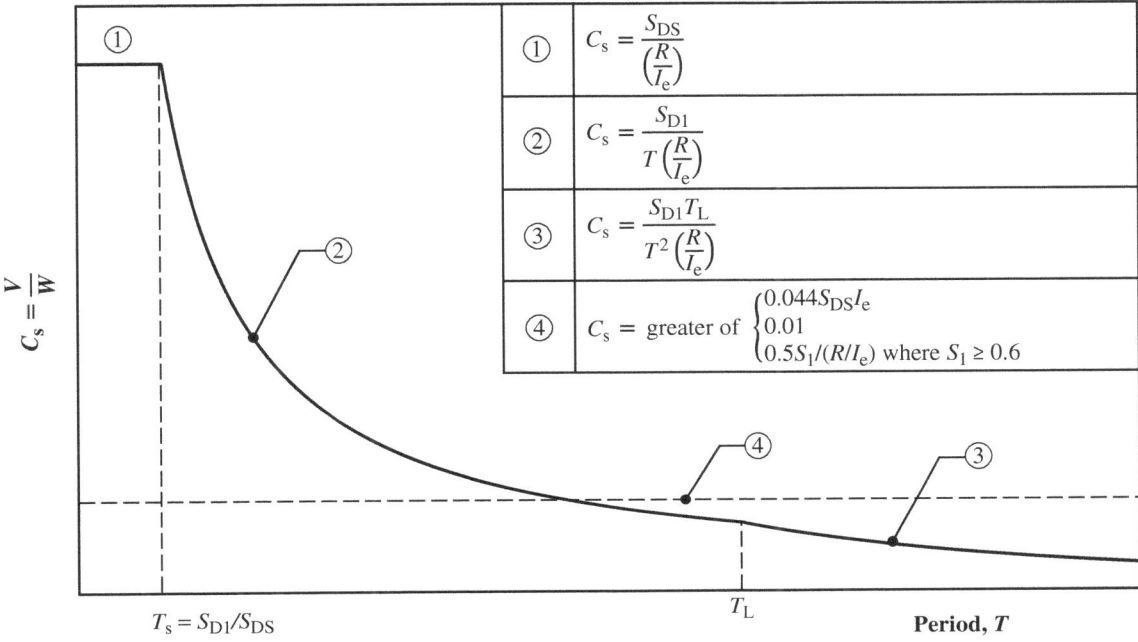

FIGURE 3.10   Design response spectrum in accordance with the ELF procedure.

Equations to determine the seismic response coefficient, C_s, are given in ASCE/SEI 12.8.1.1, which form the design response spectrum (see Fig. 3.10). It is evident from Fig. 3.10 that the transition to the design acceleration S_{DS} in the segment from $T = 0$ to $T = T_o$ in the design spectrum of ASCE/SEI 11.4.6 (see Fig. 2.4) is not used in the ELF procedure; instead, for simplicity, the horizontal portion of the design spectrum associated with constant acceleration extends from $T = 0$ to $T = T_S$.

The lower limits of C_s in ASCE/SEI Equations (12.8-5) and (12.8-6) are also indicated in Fig. 3.10. ASCE/SEI Equation (12.8-5) provides a minimum base shear as a function of S_{DS} and I_e. ASCE/SEI Equation (12.8-6) is applicable for structures located where $S_1 \geq 0.6$, which applies to sites near active faults.

The seismic response coefficient, C_s, and the vertical seismic load effect, E_v, are permitted to be calculated using S_{DS} equal to the greater of 1.0 or $0.7S_{DS}$ provide all the criteria in ASCE/SEI 12.8.1.3 are satisfied. This cap on S_{DS} reflects engineering judgement on the performance of regular, low-rise buildings in past earthquakes that comply to code-prescribed design and detailing requirements.

3.9.3   Period Determination

The fundamental period of a structure, T, in the direction of analysis must be determined based on the structural properties and deformational characteristics of the SFRS using fundamental principles of a dynamic analysis (ASCE/SEI 12.8.2).

In the preliminary design stage, some of the information needed to determine T from a dynamic analysis may not be known. ASCE/SEI Equations (12.8-7) through

(12.8-10) can be used to determine an approximate fundamental period, T_a, for a variety of structure types (see Table 3.14). These equations provide a lower-bound value for the approximate period, which, in turn, provides an upper-bound value of V.

The fundamental period can also be calculated by a first-order Rayleigh analysis where T is determined using the displacements δ_{xe} produced by the lateral forces F_i:

$$T = 2\pi \sqrt{\frac{\sum_{i=1}^{n} \delta_{xe}^2 w_i}{g \sum_{i=1}^{n} \delta_{xe} F_i}} \qquad (3.11)$$

Structure Type		T_a
Moment-resisting frame systems in which the frames resist 100 percent of the required seismic forces and are not enclosed or adjoined by components that are more rigid and will prevent the frames from deflecting when subjected to seismic forces	Steel	$0.028h_n^{0.8}$ (h_n in ft) $0.0724h_n^{0.8}$ (h_n in m)
	Concrete	$0.016h_n^{0.9}$ (h_n in ft) $0.0466h_n^{0.9}$ (h_n in m)
Steel eccentrically braced frame in accordance with ASCE/SEI Table 12.2-1 lines B1 or D1		$0.030h_n^{0.75}$ (h_n in ft) $0.0731h_n^{0.75}$ (h_n in m)
Steel buckling-restrained braced frames		$0.030h_n^{0.75}$ (h_n in ft) $0.0731h_n^{0.75}$ (h_n in m)
Structures less than or equal to 12 stories in height where the SFRS consists entirely of concrete or steel moment-resisting frames where the average story height is greater than or equal to 10 ft (3.1 m)		$0.1\,N$
Masonry or concrete shear wall structures		$\dfrac{C_q h_n}{\sqrt{C_w}}$ where $C_q = 0.0019$ ft (0.0058 m) $C_w = \dfrac{100}{A_B} \displaystyle\sum_{i=1}^{x} \dfrac{A_i}{\left[1 + 0.83\left(\dfrac{h_n}{D_i}\right)^2\right]}$
All other structural systems		$0.020h_n^{0.75}$ (h_n in ft) $0.0488h_n^{0.75}$ (h_n in m)

h_n = vertical distance from the base to the highest level of the SFRS of a structure
N = number of stories above the base of a structure
A_B = area of base of structure, ft^2 (m^2)
A_i = web area of shear wall i, ft^2 (m^2)
D_i = length of shear wall i, ft (m)
x = number of shear walls in the building effective in resisting lateral forces in the direction of analysis

TABLE 3.14 Approximate Period, T_a

The forces F_i can be determined using ASCE/SEI Equations (12.8-11) and (12.8-12), which are based on the exponent k, which in turn, is based on T (see Sec. 3.9.4 of this publication). Thus, to use this method, it is recommended that k be determined assuming T is equal to the upper limit prescribed in ASCE/SEI 12.8.2 (see below). Once T is calculated by Eq. (3.11), it can be used to obtain an updated k, an updated set of F_i and δ_{xe}, and a more refined T.

The calculated fundamental period, T, must be taken less than or equal to $C_u T_a$ where C_u is the coefficient for upper limit on calculated period given in ASCE/SEI Table 12.8-1 as a function of S_{D1}. The purpose of setting a limit on the calculated period is to ensure that an unusually low-base shear is not obtained for overly flexible structures.

A summary of the periods to be used when calculating the lateral forces on a structure is as follows:

- Where the calculated $T \le T_a$: use $T = T_a$
- Where $T_a <$ calculated $T < C_u T_a$: use $T =$ calculated T
- Where the calculated $T \ge C_u T_a$: use $T = C_u T_a$

3.9.4   Vertical Distribution of Seismic Forces

The seismic base shear, V, is distributed over the height of the building in accordance with ASCE/SEI Equations (12.8-11) and (12.8-12) where F_x is the lateral seismic force at level x of the building (see Fig. 3.11). The parameters in these equations are as follows:

- w_i and w_x are the portions of W located or assigned to level i or x
- h_i and h_x are the heights from the base of the structure to level i or x
- k is the exponent related to the structure period:

$$k = \begin{cases} 1.0 \text{ for } T \le 0.5 \ s \\ 0.75 + 0.5T \text{ for } 0.5 \ s < T < 2.5 \ s \\ 2.0 \text{ for } T \ge 2.5 \ s \end{cases}$$

The parameter k accounts for higher mode effects, and it is permitted to take $k = 2.0$ for structures with a period between 0.5 and 2.5 s instead of using the above equation for that period range.

3.9.5   Horizontal Distribution of Forces

Overview
The seismic design story shear, V_x, in story x is the sum of the lateral force, F_i, acting at the floor level supported by that story plus the sum of the lateral forces of all the floor levels above, ΣF_i, including the roof. The story shear is distributed to the vertical elements of the SFRS based on the in-plane stiffness of the diaphragm.

Flexible Diaphragms
For flexible diaphragms, it is assumed the horizontal seismic forces are distributed to the vertical elements of the SFRS in proportion to the mass tributary to the vertical elements (see Table 3.4). For diaphragms of uniform thickness and material, the horizontal force allocation is in proportion to the areas tributary to the vertical elements. Flexible

82

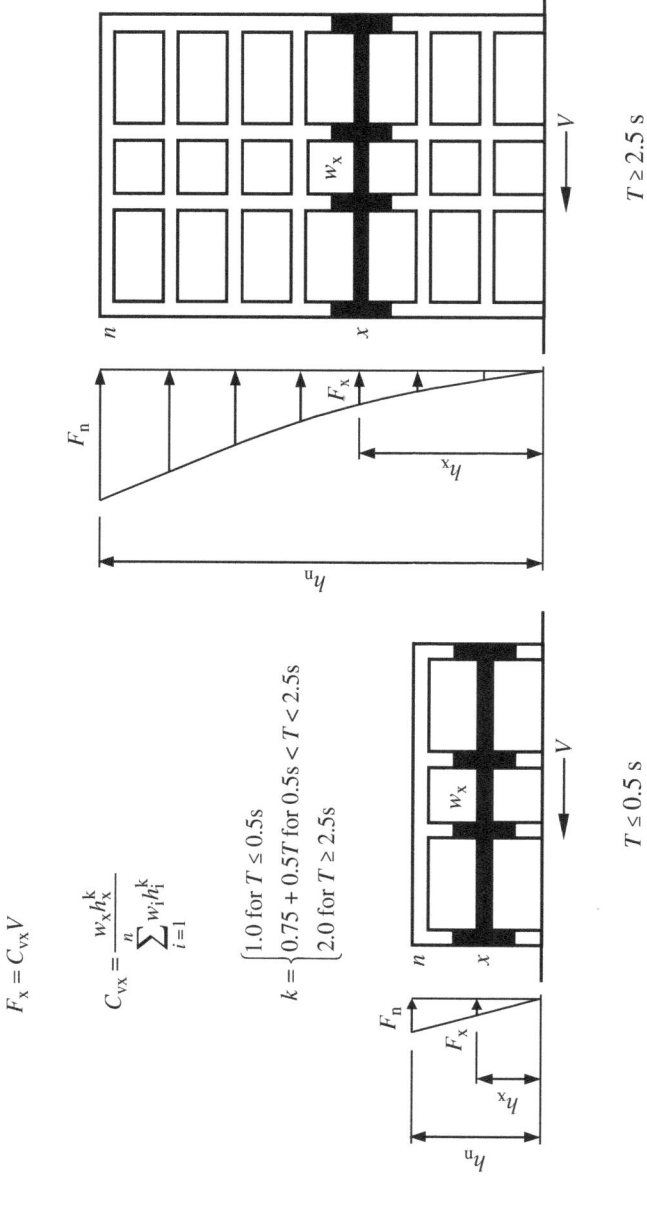

$$F_x = C_{vx}V$$

$$C_{vx} = \frac{w_x h_x^k}{\displaystyle\sum_{i=1}^{n} w_i h_i^k}$$

$$k = \begin{cases} 1.0 \text{ for } T \le 0.5\text{s} \\ 0.75 + 0.5T \text{ for } 0.5\text{s} < T < 2.5\text{s} \\ 2.0 \text{ for } T \ge 2.5\text{s} \end{cases}$$

Figure 3.11 Vertical distribution of seismic forces (ASCE/SEI 12.8.3).

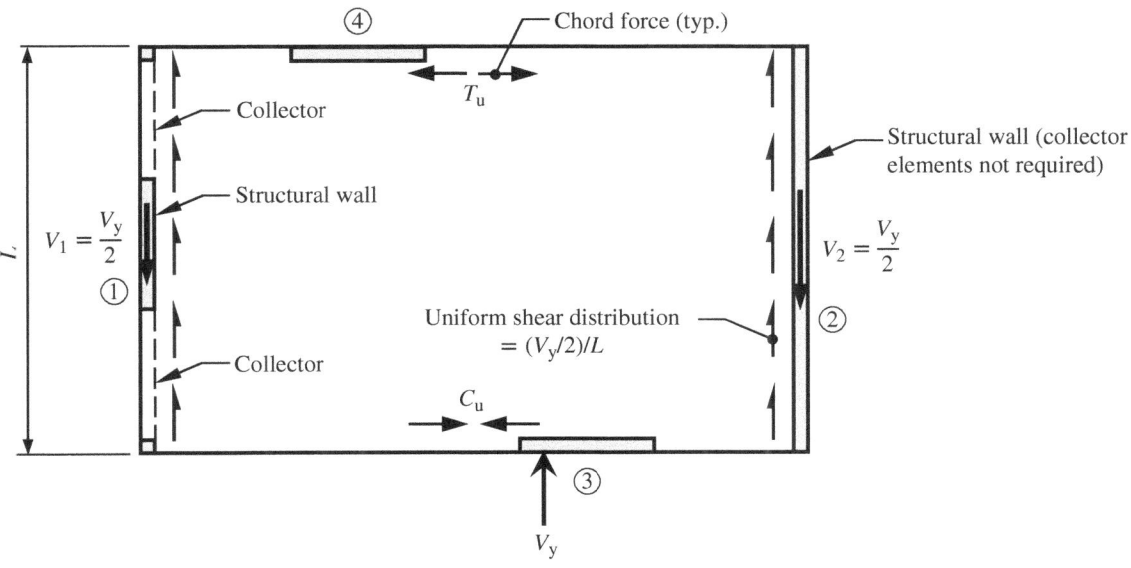

Figure 3.12 Horizontal distribution of forces in a flexible diaphragm.

diaphragms are not capable of distributing horizontal seismic forces in proportion to the rigidities (or, stiffnesses) of the vertical elements of the SFRS. Also, torsion generated by the application of the horizontal seismic forces need not be considered in flexible diaphragms.

Consider the flexible diaphragm in Fig. 3.12. The horizontal seismic force, V_y, is transferred through the web of the diaphragm to the shear walls, which act as supports for the diaphragm. Walls 1 and 2 each resist 50 percent of the seismic force. A uniform shear distribution occurs along the depth of the diaphragm adjacent to the walls and is equal to $(V_y/2)/L$. Collectors are required at wall 1 because the wall does not extend the full depth of the diaphragm.

The chords at the top and bottom edges of the diaphragm are perpendicular to the direction of the seismic force and resist the tension and compression forces induced in the diaphragm due to bending.

Rigid Diaphragms

For rigid diaphragms, horizontal forces are distributed to the vertical elements of the SFRS in proportion to their relative rigidities and their locations with respect to the center of rigidity (CR). Thus, the location of the CR must be determined on a floor/roof level prior to horizontal force allocation. By definition, the CR is the point on a floor/roof level where the equivalent story stiffness is assumed to be located. It is often referred to as the stiffness centroid. For buildings with rigid diaphragms, application of a lateral force through the CR produces only rigid body displacement of the story. Displacement and rotation occur where the lateral force is applied at any point other than the CR. Depending on the structural layout, the CR can be at different locations on different levels in a building.

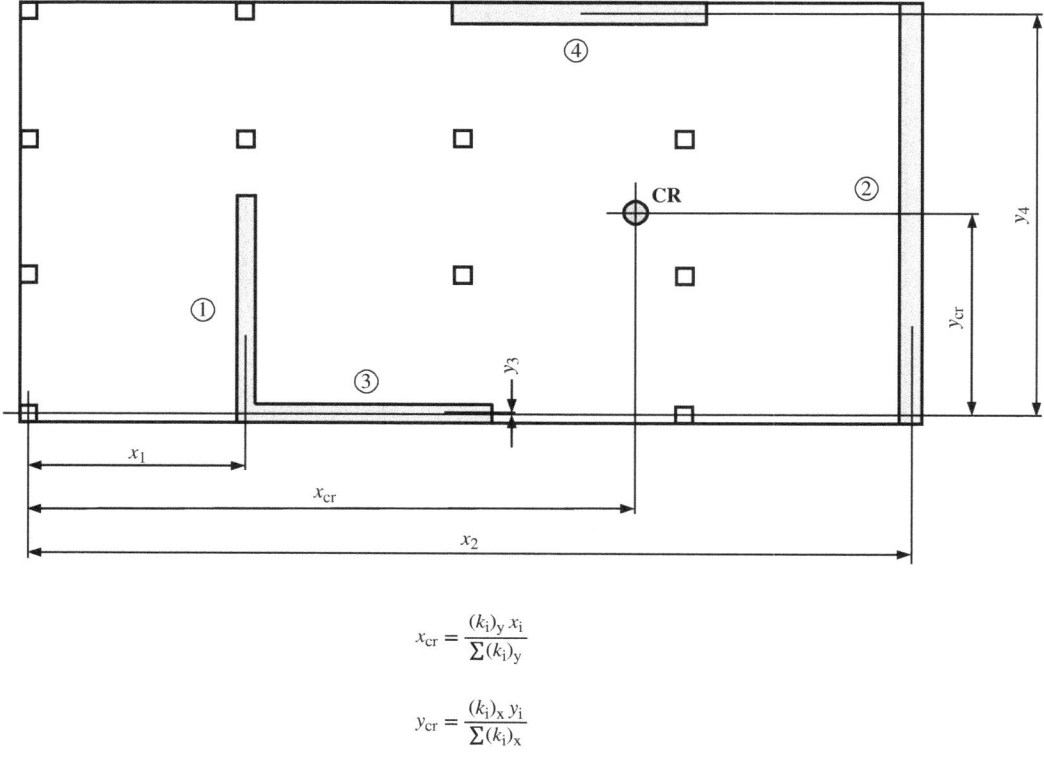

$$x_{cr} = \frac{(k_i)_y \, x_i}{\sum (k_i)_y}$$

$$y_{cr} = \frac{(k_i)_x \, y_i}{\sum (k_i)_x}$$

Figure 3.13 Determination of the location of the CR.

The location of the CR in the x-direction (x_{cr}) and in the y-direction (y_{cr}) can be determined using the equations in Fig. 3.13. The terms $(k_i)_y$ and $(k_i)_x$ are the in-plane stiffnesses of lateral force–resisting elements i in the y- and x-directions, respectively, and x_i and y_i are the distances in the x- and y-directions from an arbitrarily selected origin to the centroids of the lateral force–resisting elements i. All elements of the SFRS parallel to direction of analysis are included in the equations to determine the location of the CR; it is commonly assumed that out-of-plane resistance of a lateral force–resisting element is negligible compared to its in-plane resistance. Thus, when determining the location of the CR in the x-direction, only the stiffnesses of walls 1 and 2 are considered. A portion of wall 3 (that is, an effective flange width) may be included when determining the stiffness of wall 1. Similarly, only the stiffness of wall 3 (including an effective flange width of wall 1, if desired) and wall 4 are considered when determining the location of the CR in the y-direction.

In-plane stiffness, k_i, can be obtained by any rational method. Approximate in-plane stiffnesses of solid walls, rigid frames, and braced frames are given in Fig. 3.14.

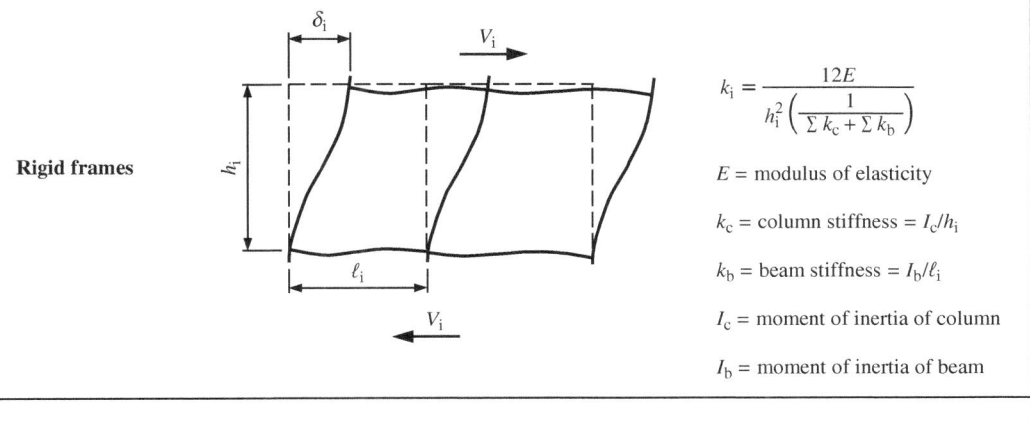

Rigid frames

$$k_i = \dfrac{12E}{h_i^2 \left(\dfrac{1}{\Sigma k_c + \Sigma k_b} \right)}$$

E = modulus of elasticity

k_c = column stiffness = I_c/h_i

k_b = beam stiffness = I_b/ℓ_i

I_c = moment of inertia of column

I_b = moment of inertia of beam

Solid walls or piers

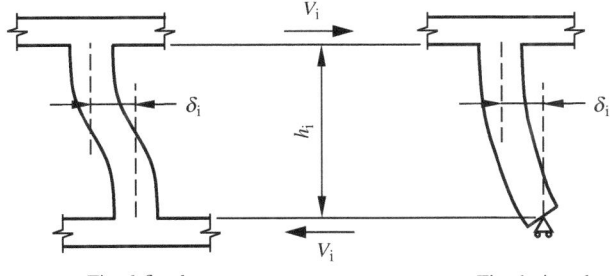

Fixed-fixed   Fixed-pinned

$$k_i = \dfrac{k_{Fi} k_{Vi}}{k_{Fi} + k_{Vi}}$$

E = modulus of elasticity

G = shear modulus

k_{Fi} = flexural stiffness = V_i/δ_{Fi}

k_{Vi} = shear stiffness = V_i/δ_{Vi}

I_i = moment of inertia of wall or pier

A_i = area of wall or pier

Support condition	Flexural deflection, δ_{Fi}	Shear deflection, δ_{Vi}
Fixed-fixed	$\dfrac{V_i h_i^3}{12EI_i}$	$\dfrac{1.2 V_i h_i}{GA_i}$
Fixed-pinned	$\dfrac{V_i h_i^3}{3EI_i}$	$\dfrac{1.2 V_i h_i}{GA_i}$

FIGURE 3.14 Approximate in-plane stiffnesses of solid walls, rigid frames, and braced frames.

Braced Bents		
Single diagonal	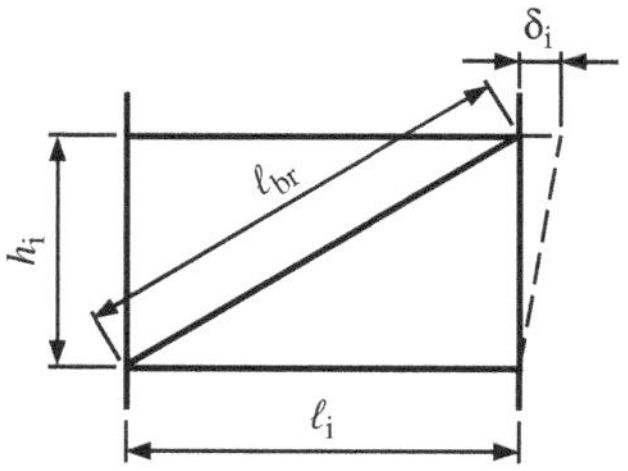	$$k_i = \dfrac{E}{\left(\dfrac{\ell_{br}^3}{A_{br}\ell_i^2} + \dfrac{\ell_i}{A_b}\right)}$$ E = modulus of elasticity A_{br} = area of brace A_b = area of upper beam
Double diagonal	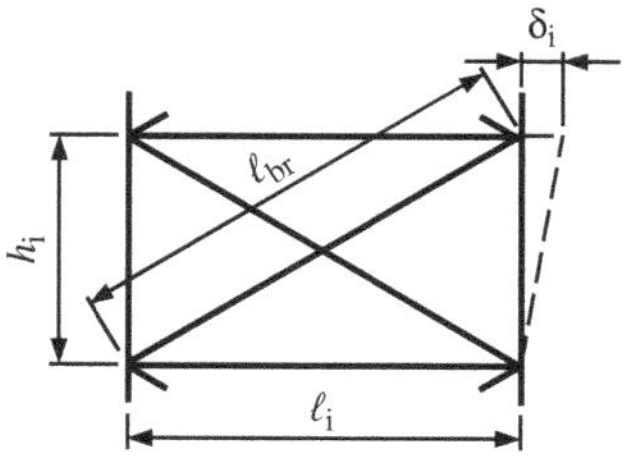	$$k_i = \dfrac{2E}{\left(\dfrac{\ell_{br}^3}{A_{br}\ell_i^2}\right)}$$ E = modulus of elasticity A_{br} = area of brace
K-brace	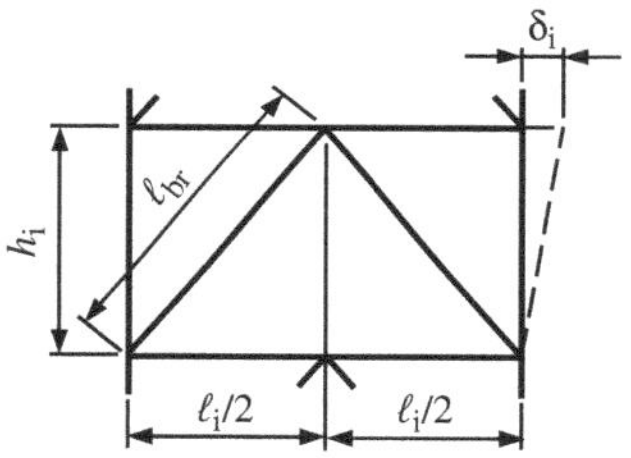	$$k_i = \dfrac{E}{\left(\dfrac{2\ell_{br}^3}{A_{br}\ell_i^2} + \dfrac{\ell_i}{4A_b}\right)}$$ E = modulus of elasticity A_{br} = area of brace A_b = area of upper beam
Knee brace	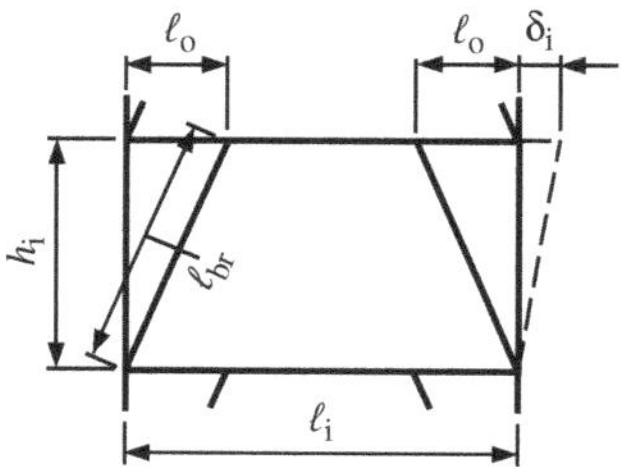	$$k_i = \dfrac{E}{\left[\dfrac{\ell_{br}^3}{2A_{br}\ell_o^2} + \dfrac{\ell_o}{2A_b} + \dfrac{h_i^2(\ell_i - 2\ell_o)^2}{12I_b\ell_i}\right]}$$ E = modulus of elasticity A_{br} = area of brace A_b = area of upper beam I_b = moment of inertia of upper beam
Offset diagonal	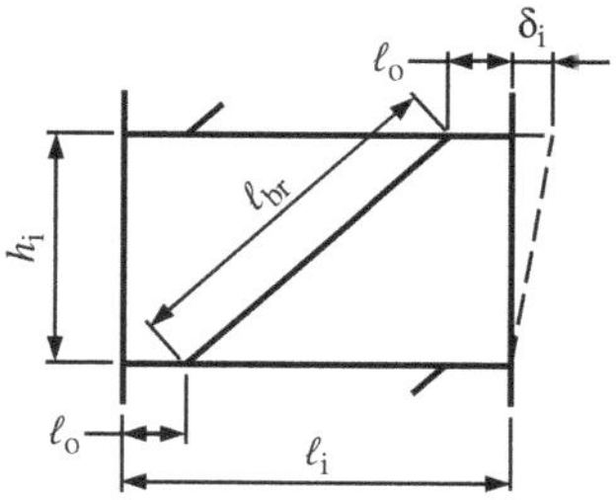	$$k_i = \dfrac{E}{\left[\dfrac{\ell_{br}^3}{A_{br}(\ell_i - 2\ell_o)^2} + \dfrac{\ell_i - 2\ell_o}{A_b} + \dfrac{h_i^2\ell_o^2}{3I_b\ell_i}\right]}$$ E = modulus of elasticity A_{br} = area of brace A_b = area of upper beam I_b = moment of inertia of upper beam

FIGURE 3.14   *(Continued)*

The center of mass (CM) is the location on a level of a building where the mass of the entire story is assumed to be concentrated. The seismic force for the story is assumed to act at this location.

The equations in Fig. 3.15 can be used to determine the portion of the total story shear, V_y, in the y-direction that is resisted by element i of the SFRS considering both direct and torsional shear forces, which is designated $(V_i)_y$. In these equations, $\bar{x}_i$ = perpendicular distance from the centroid of element i to the CR parallel to the x-axis and $\bar{y}_i$ = perpendicular distance from the centroid of element i to the CR parallel to the y-axis.

Term 1 is the portion of the total story shear resisted by element i based on relative stiffness. Only the elements of the MWFRS parallel to the y-direction share this force (see Fig. 3.15, which indicates term 1 forces for walls 1 and 2 in the direction resisting the applied seismic force, V_y). Term 2 is the shear force resisted by element i that is generated by the inherent torsional moment, $M_t = V_y e_x$ (ASCE/SEI 12.8.4.1). The elements of the MWFRS in both directions share the shear forces generated by this moment (see Fig. 3.15, which indicates term 2 forces for all four walls in the directions resisting the applied clockwise torsional moment). The forces in walls 3 and 4 are calculated using term 2 with the numerator equal to $\bar{y}_i(k_i)_x$ instead of $\bar{x}_i(k_i)_y$. The denominator of this term is the torsional stiffness of all the walls and is analogous to the polar moment of inertia of a section. Term 2 is equal to zero in this equation when V_y acts through the CR; in such cases, the floor translates as a rigid body and the elements of the MWFRS all displace an equal amount horizontally.

Term 2 forces may be either positive or negative depending on the location of the element with respect to the CR. For example, for seismic forces in the y-direction, the term 2 force for wall 2 acts in the opposite direction of the term 1 force (see Fig. 3.15). To properly capture the maximum force effects on an element, seismic forces acting in the opposite direction must also be considered.

Equations for horizontal force distribution in the x-direction for seismic story shear V_x are similar to those in Fig. 3.15.

The accidental torsional moment, M_{ta}, must also be included for diaphragms that are not flexible when considering whether a horizontal irregularity in ASCE/SEI Table 12.3-1 exists or not. Accidental torsional moments are determined based on the assumption that the CM is displaced each way from its actual location by a distance equal to 5 percent of the dimension of the structure perpendicular to the direction of analysis.

Illustrated in Fig. 3.16 is a floor or roof diaphragm that is not flexible. The seismic story shear, V_y, acts through the CM and the inherent torsional moment, M_t, is equal to $V_y e_x$ where e_x is the eccentricity between the CM and the CR perpendicular to the direction of analysis. The accidental torsional moment, M_{ta}, is equal to $V_y(0.05B)$, and the total torsional moment for purposes of analysis is equal to $M_t + M_{ta}$. Similar calculations can be performed for the seismic story shear, V_x, in the perpendicular direction using the eccentricity e_y.

In cases where forces are applied concurrently in orthogonal directions, the required 5 percent displacement of the CM need not be applied in both of the orthogonal directions simultaneously; rather, it must be applied in the direction producing the greatest effects for each element considered.

Accidental torsional moments need not be included when determining earthquake load effects, E, in the design of the structure (including the diaphragms) and in the determination of the design story drifts in accordance with ASCE/SEI 12.8.6, 12.9.1.2,

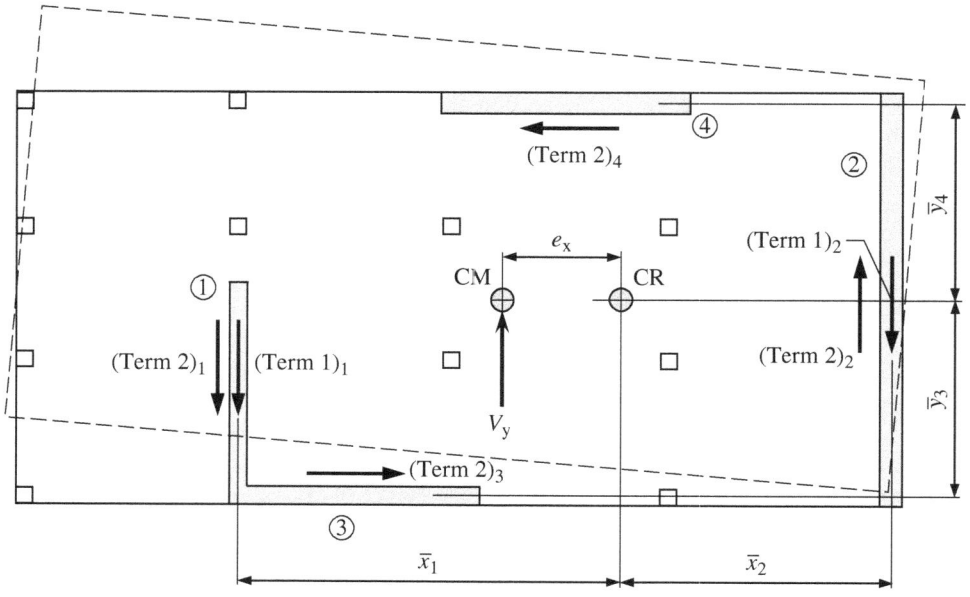

$$(V_i)_y = \frac{(k_i)_y}{\Sigma\,(k_i)_y}V_y \pm \frac{\overline{x}_i(k_i)_y}{\Sigma\,\overline{x}_i^2\,(k_i)_y + \Sigma\,\overline{y}_i^2\,(k_i)_x}V_y e_x$$

$$\underbrace{\phantom{\frac{(k_i)_y}{\Sigma\,(k_i)_y}}}_{\text{Term 1}} \qquad \underbrace{\phantom{\frac{\overline{x}_i(k_i)_y}{\Sigma\,\overline{x}_i^2\,(k_i)_y}}}_{\text{Term 2}}$$

$$(V_1)_y = \frac{(k_1)_y}{(k_1)_y + (k_2)_y}V_y + \frac{\overline{x}_1(k_1)_y}{\overline{x}_1^2\,(k_1)_y + \overline{x}_2^2\,(k_2)_y + \overline{y}_3^2\,(k_3)_x + \overline{y}_4^2\,(k_4)_x}V_y e_x$$

$$(V_2)_y = \frac{(k_2)_y}{(k_1)_y + (k_2)_y}V_y - \frac{\overline{x}_2(k_2)_y}{\overline{x}_1^2\,(k_1)_y + \overline{x}_2^2\,(k_2)_y + \overline{y}_3^2\,(k_3)_x + \overline{y}_4^2\,(k_4)_x}V_y e_x$$

$$(V_3)_y = \frac{\overline{y}_3(k_3)_x}{\overline{x}_1^2\,(k_1)_y + \overline{x}_2^2\,(k_2)_y + \overline{y}_3^2\,(k_3)_x + \overline{y}_4^2\,(k_4)_x}V_y e_x$$

$$(V_4)_y = \frac{\overline{y}_4(k_4)_x}{\overline{x}_1^2\,(k_1)_y + \overline{x}_2^2\,(k_2)_y + \overline{y}_3^2\,(k_3)_x + \overline{y}_4^2\,(k_4)_x}V_y e_x$$

FIGURE 3.15 Horizontal distribution of forces in a rigid diaphragm.

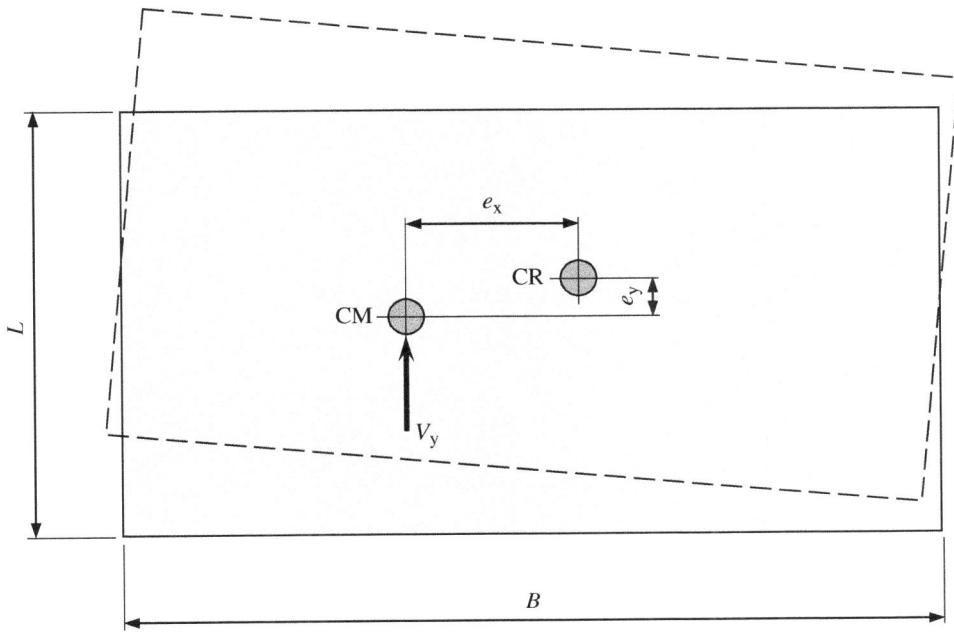

Inherent torsional moment: $M_t = V_y e_x$

Accidental torsional moment: $M_{ta} = V_y(0.05B)*$

Total torsional moment: $M_t + M_{ta}$

$*M_{ta}$ to be included only when required by ASCE/SEI 12.8.4.2

Figure 3.16 Inherent and accidental torsional moments.

or Chapter 16, including the story drift limits in ASCE/SEI 12.12.1, except for the following structures (see Table 3.6):

- Structures assigned to SDC B with type 1b horizontal structural irregularity.
- Structures assigned to SDC C, D, E, or F with type 1a or type 1b horizontal structural irregularity.

The accidental torsional moment must be amplified in accordance with ASCE/SEI 12.8.4.3 in structures assigned to SDC C, D, E, or F where a type 1a or type 1b torsional irregularity is present. At each level, M_{ta} must be multiplied by the torsional amplification factor, A_x, which is determined by ASCE/SEI Equation (12.8-14):

$$1.0 \le A_x = \left(\frac{\delta_{max}}{1.2\delta_{avg}} \right)^2 \le 3.0 \tag{3.12}$$

In this equation, δ_{max} is the maximum displacement that occurs at level x in the structure determined assuming $A_x = 1.0$ and δ_{avg} is the average of the displacements at

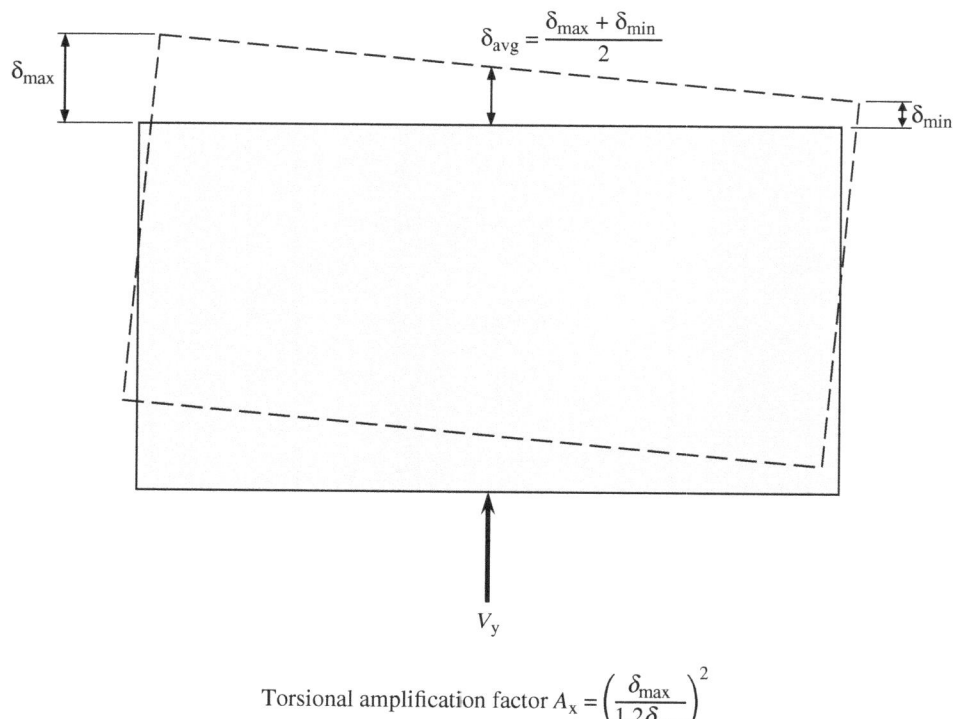

$$\delta_{avg} = \frac{\delta_{max} + \delta_{min}}{2}$$

$$\text{Torsional amplification factor } A_x = \left(\frac{\delta_{max}}{1.2\delta_{avg}}\right)^2$$

FIGURE 3.17   Torsional amplification factor, A_x.

the extreme points of the structure at level x assuming $A_x = 1.0$ (see Fig. 3.17). Setting $A_x = 1.0$ when determining the displacements eliminates the need for iterations in the calculation of A_x.

3.9.6   Overturning

The structure must be designed to resist overturning effects caused by the seismic forces (ASCE/SEI 12.8.5). The critical load combinations are typically those where the effects from gravity and seismic loads counteract.

3.9.7   Story Drift Determination

Design story drift, Δ, is determined in accordance with ASCE/SEI 12.8.6 and is defined as the difference of the deflections, δ_x, at the CM of the diaphragms at the top and bottom of the story under consideration (see Fig. 3.18).

The horizontal displacements at each level due to the application of the seismic forces F_x are designated δ_{xe} and are obtained from an elastic analysis of the structure. The deflections δ_x are equal to the following (see ASCE/SEI Equation (12.8-15)):

$$\delta_x = \frac{C_d \delta_{xe}}{I_e} \tag{3.13}$$

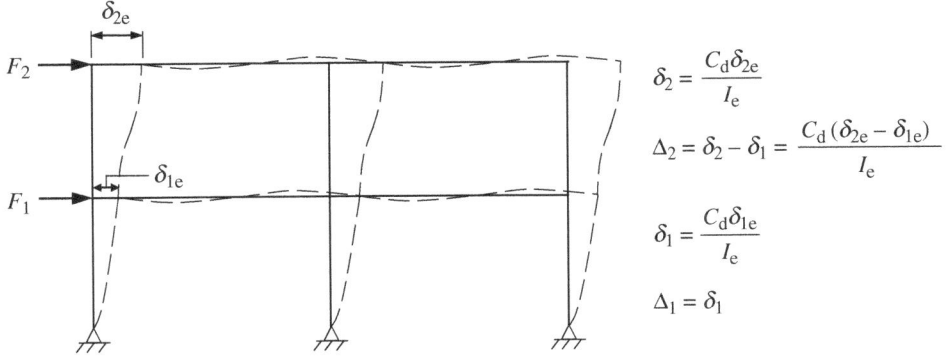

FIGURE 3.18   Story drift determination in accordance with ASCE/SEI 12.8.6.

The elastic deflections, δ_{xe}, are multiplied by the deflection amplification factor, C_d, of the SFRS to estimate the deflections likely to occur from the design earthquake ground motion (see Fig. 3.1). Because the seismic forces F_x are determined using the importance factor, I_e, the elastic deflections, δ_{xe}, are divided by I_e so that the calculated deflections δ_x are not overly conservative in cases where I_e is greater than 1.0.

The minimum seismic base shear calculated in accordance with ASCE/SEI Equation (12.8-5) need not be considered when determining drifts (ASCE/SEI 12.8.6.1).

The computed drifts must be less than or equal to the story drift limits in ASCE/SEI 12.12.1 (see Sec. 3.13 of this publication). For determining compliance with the allowable story drifts, it is permitted to determine δ_{xe} using seismic design forces based on the computed fundamental period of the structure without the upper limit of $C_u T_a$ in ASCE/SEI 12.8.2 (ASCE/SEI 12.8.6.2).

3.9.8   P-Delta Effects

As a structure deflects horizontally due to seismic forces, the vertical loads are displaced from their original position and additional effects are introduced into the structural members (that is, P-delta effects), which cause the structure to displace further. If sufficient strength and stiffness are not provided, the deflections will continue to increase, leading to overall instability of the structure.

Member forces and story drifts induced by P-delta effects must be considered in member design where such effects are significant (ASCE/SEI 12.8.7). In lieu of a more refined analysis, P-delta effects need not be considered where the stability coefficient, θ, is less than or equal to 0.10 where θ is determined by ASCE/SEI Equation (12.8-16) (see Fig. 3.19):

$$\theta = \frac{P_x \Delta I_e}{V_x h_{sx} C_d} \tag{3.14}$$

where
  P_x = the total vertical design load above level $x = \sum_{i=x}^{n} (P_D + P_L)_i$
  P_D = vertical unfactored dead loads

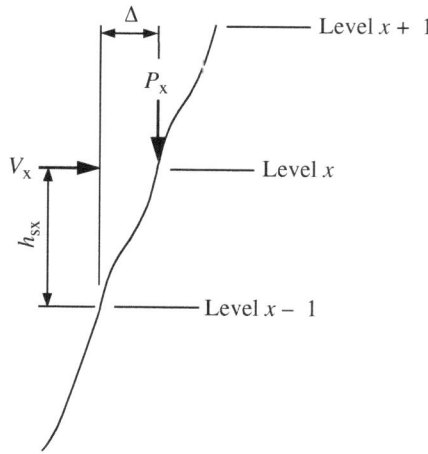

FIGURE 3.19 P-delta effects.

P_L = vertical unfactored, reduced live loads

n = total number of levels in the structure

Δ = design story drift defined in ASCE/SEI 12.8.6 occurring simultaneously with V_x (see Sec. 3.9.7 of this publication)

V_x = seismic force acting between levels x and $x - 1$

h_{sx} = story height below level x

A structure is considered to be potentially unstable when the following equation is satisfied (see ASCE/SEI (12.8-17)):

$$\theta > \theta_{max} = \frac{0.5}{\beta C_d} \leq 0.25 \qquad (3.15)$$

The term β is the ratio of the shear demand to the shear capacity of the story under consideration, which is permitted to be conservatively taken as 1.0. Equation (3.15) must be checked even where an automated analysis is used to determine the P-delta effects. However, the value of θ calculated by Eq. (3.14) is permitted to be divided by $(1+\theta)$ prior to checking Eq. (3.15).

Where P-delta effects must be considered (that is, $0.10 < \theta \leq \theta_{max}$), the displacement and member forces must be increased accordingly using a rational analysis. Alternatively, it is permitted to multiply displacements and member forces by $1.0/(1-\theta)$. Rational analysis methods are presented in Ref. 3.

The flowchart in Fig. 3.20 can be used to determine seismic base shear and its vertical distribution on an SFRS of a structure in accordance with the ELF procedure.

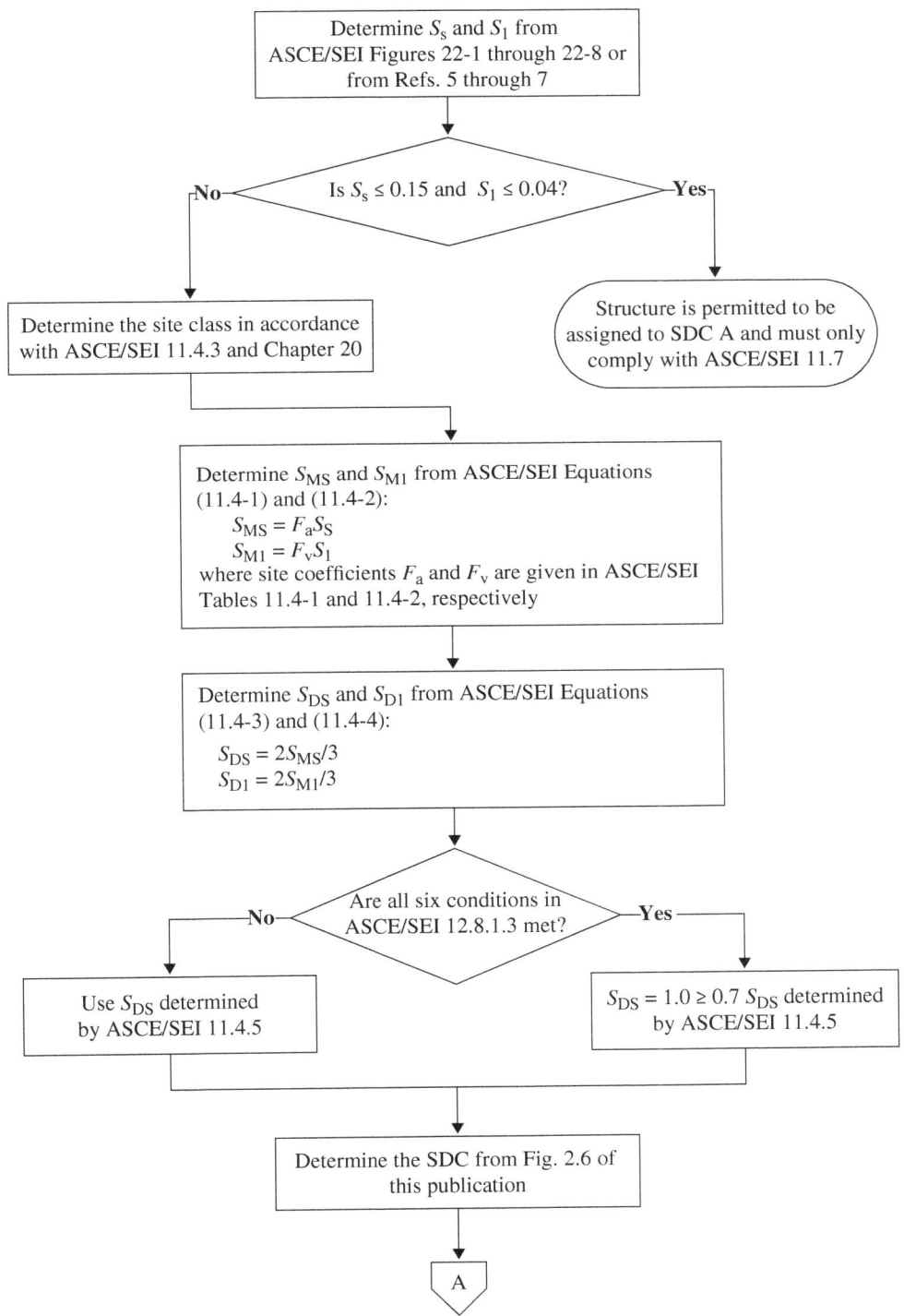

FIGURE 3.20 Determination and distribution of seismic forces in accordance with the ELF procedure.

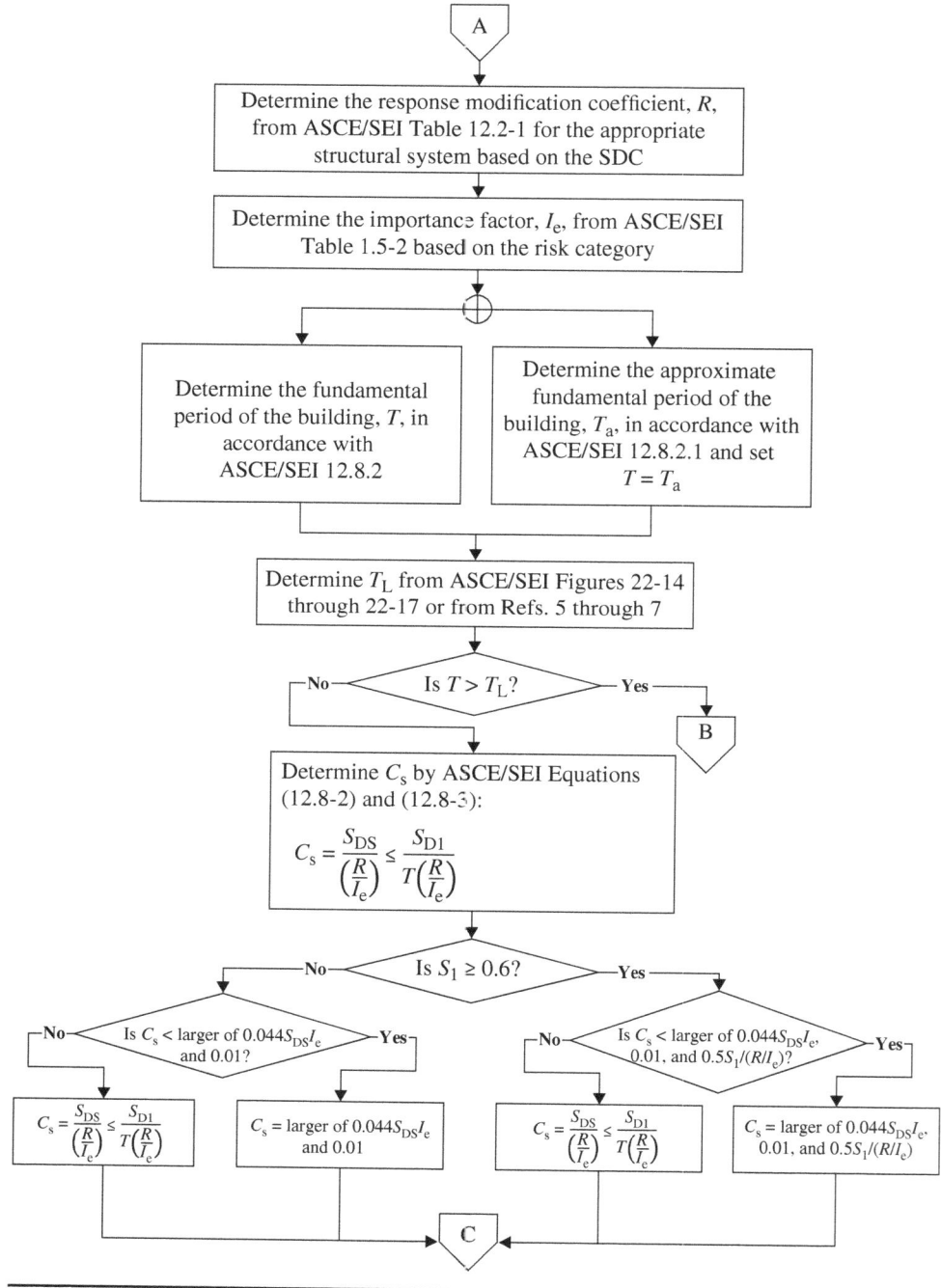

FIGURE 3.20 *(Continued)*

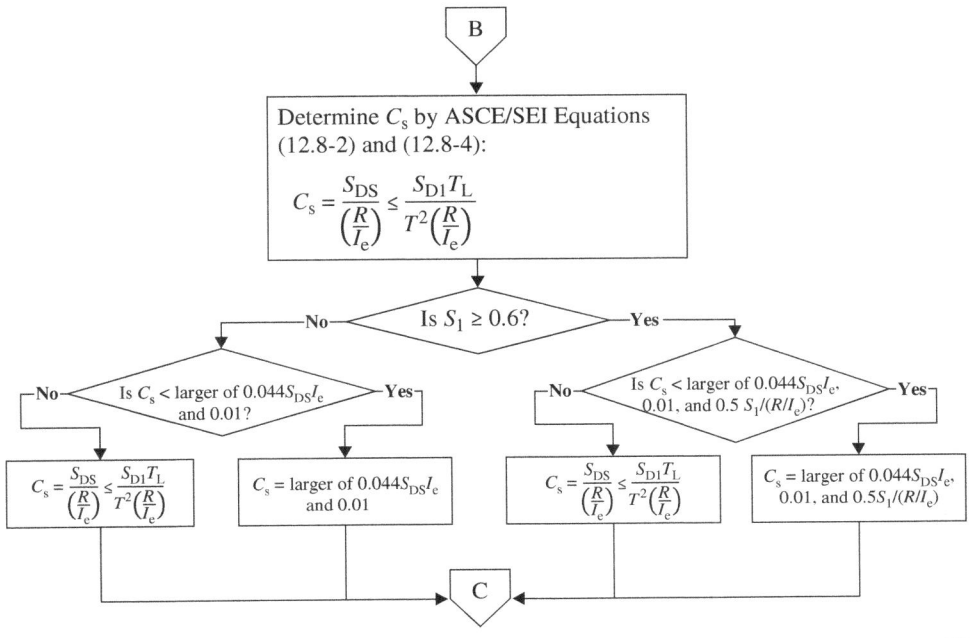

FIGURE 3.20   *(Continued)*

3.10   Linear Dynamic Analysis

3.10.1   Modal Response Spectrum Analysis

In the modal response spectrum analysis, a structure is broken down into a number of single-degree-of-freedom systems. Each system has its own mode shape and natural period of vibration.

For seismic design, the number of modes corresponds to the number of mass degrees of freedom of the structure. Even for a simple low-rise building, the number of mass degrees of freedom is immense. Fortunately, some simple idealizations can be made to significantly reduce the number of degrees of freedom while still preserving a meaningful solution. For example, in the case of planar structures with rigid diaphragms, the number of mass degrees of freedom can be taken as one per story (horizontal translation in the direction of analysis) with the mass of the story concentrated in the floor and roof systems.

The three modes of vibration for a three-story building are illustrated in Fig. 3.21. For three-dimensional structures with rigid diaphragms, three mass degrees of freedom can be utilized per story (two horizontal translations and rotation about the vertical axis). The number of modes that is required for design is discussed below.

Details on how to determine modal periods, mode shapes, and modal participation are given in numerous standard references on dynamic analysis and are not covered here (see for example, Ref. 9).

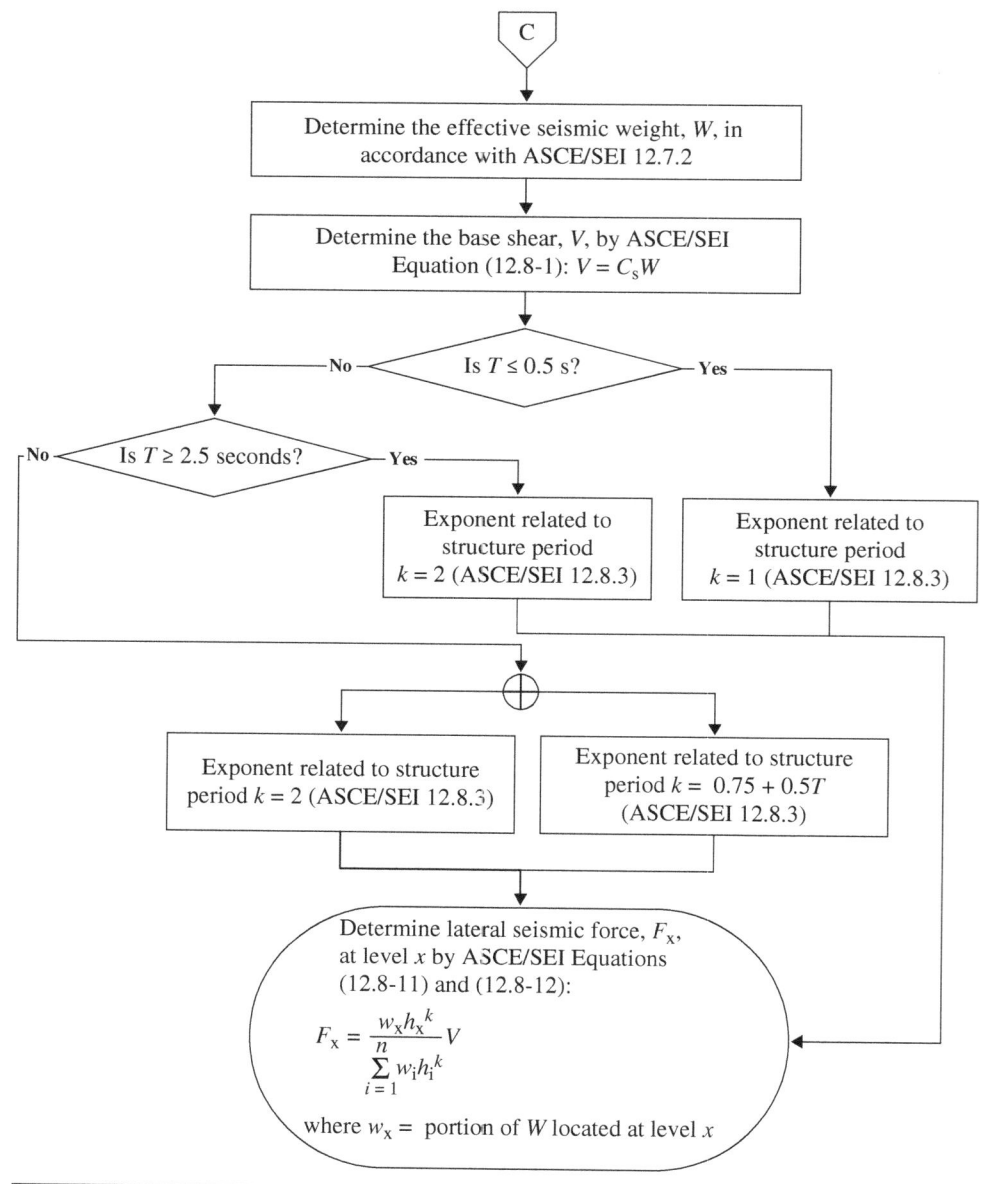

FIGURE 3.20   *(Continued)*

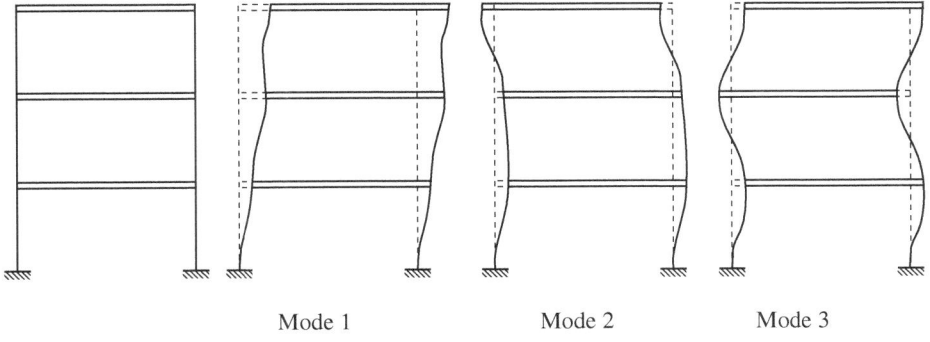

Mode 1   Mode 2   Mode 3

FIGURE 3.21 Modes of vibration for an idealized planar structure.

A modal response spectrum analysis is permitted for any structure assigned to SDC B through F and is required for regular and irregular structures where $T \geq 3.5T_S$ and certain types of irregular structures assigned to SDC D through F (see Table 3.12). The requirements in Table 3.15 must be satisfied for this method of analysis (ASCE/SEI 12.9.1).

3.10.2   Linear Response History Analysis

Requirements for the linear response history analysis method are given in ASCE/SEI 12.9.2. In this method, which is considered an alternate to the modal response spectrum analysis method, a three-dimensional model of a structure is constructed in accordance with ASCE/SEI 12.9.2.2, and the model is subjected to suites of spectrally matched acceleration histories compatible with the design response spectrum for the site. The system response is obtained by simultaneous solution of the full set of equations of motion. ASCE/SEI C12.9.2 gives additional information on the proper application of this method.

3.11   Diaphragms, Chords, and Collectors

3.11.1   Overview

A diaphragm is a horizontal or sloped system that transmits lateral forces to the vertical elements of the SFRS. Horizontal seismic forces are transferred to the elements of the SFRS based on the flexibility of the diaphragm (see Sec. 3.9.5 of this publication).

Diaphragms also transfer gravity loads that are perpendicular to the diaphragm surface to floor or roof members (beams, joists, and columns). When properly attached to the top surface of horizontal framing members, diaphragms increase the flexural lateral stability of such members.

Diaphragms, chords, and collectors must be designed in accordance with ASCE/SEI 12.10.1 and 12.10.2. Precast concrete diaphragms, including chords and collectors, in structures assigned to SDC C, D, E, or F must be designed in accordance with ASCE/SEI 12.10.3 (see exception 1 in ASCE/SEI 12.10). The requirements in ASCE/SEI 12.10.3 are permitted to be used for the design of precast concrete diaphragms in structures

Number of modes	• Analysis must include a sufficient number of modes to obtain a combined modal mass participation of 100 percent of the structure's mass. It is permitted to represent all modes with periods less than 0.05 s in a single rigid body mode that has a period of 0.05 s. • Alternatively, at least 90 percent of the actual mass in each orthogonal horizontal direction must participate in the response in order to obtain a distribution of forces and displacements that are sufficient for design.
Modal response parameters	• The forces and deflections are determined for each mode using the properties of each mode and either the general response spectrum defined in ASCE/SEI 11.4.6 or the site-specific response spectrum defined ASCE/SEI 21.2 (see Sec. 2.2.8 of this publication). Regardless of the spectrum that is used, the spectral ordinates must be divided by (R/I_e). • Displacement and drift quantities must be multiplied by (C_d/I_e).
Combined response parameters	The results from the analysis of each mode are to be combined using one of the following methods: • Square root of the sum of the squares (SRSS) • Complete quadratic combination (CQC) • Complete quadratic combination as modified by ASCE 4 (CQC-4) (Ref. 6.4) • Approved equivalent approach Either the CQC or CQC-4 methods must be used where closely spaced modes have significant cross-correlation of translational and torsional response.
Scaling design values of combined response	• Scaling of forces Where the combined response for the modal base shear, V_t, determined in accordance with ASCE/SEI 12.9.1.3 is less than 100 percent of the calculated base shear, V, by the ELF procedure, the forces must be scaled by the factor V/V_t. Where the calculated fundamental period, T, exceeds C_uT_a in a given direction, C_uT_a is to be used instead of T in the analysis in that direction. • Scaling of drifts Drifts must be scaled by the factor (C_sW/V_t) when $V_t < C_sW$ where C_s is determined by ASCE/SEI Equation (12.8-6).
Horizontal shear distribution	Distribution of the horizontal forces to the SFRSs must be in accordance with ASCE/SEI 12.8.4 (see Sec. 3.9.5 of this publication). Amplification of torsion in accordance with ASCE/SEI 12.8.4.3 is not required where accidental torsion effects are included in the model.
P-delta effects	The P-delta effects of ASCE/SEI 12.8.7 must be satisfied. The base shear that is used to determine the story shears and story drifts is determined in accordance with ASCE/SEI 12.8.6.
Soil-structure interaction reduction	Reduction in forces associated with soil structure interaction is permitted provided the requirements of ASCE/SEI Chapter 19 are used. Any generally accepted procedures approved by the authority having jurisdiction are also permitted.
Structural modeling	A three-dimensional mathematical model of the structure is required, which must be constructed in accordance with ASCE/SEI 12.7.3. In structures without rigid diaphragms, the model must include the stiffness characteristics of the diaphragm and any additional dynamic degrees of freedom needed to accurately account for the participation of the diaphragm in the dynamic response of the structure.

TABLE 3.15 Requirements for the Modal Response Spectrum Analysis

assigned to SDC B, cast-in-place concrete diaphragms, and wood-sheathed diaphragms supported by wood diaphragm framing.

3.11.2 Diaphragm Design

According to ASCE/SEI 12.10.1, diaphragms must be designed for the shear and bending stresses caused by the seismic design forces. The resulting shear and tension forces must be less than or equal to the corresponding design strengths of the diaphragm.

For structures assigned to SDC B and higher, floor and roof diaphragms must be designed to resist the effects from the larger of the following (ASCE/SEI 12.10.1.1):

- The lateral seismic force, F_x, determined from the vertical distribution of the seismic base shear, V (see Fig. 3.11).

- The diaphragm design force, F_{px}, determined by ASCE/SEI Equation (12.10-1), including the minimum and maximum limits in ASCE/SEI Equations (12.10-2) and (12.10-3), respectively:

$$F_{px} = \frac{\sum_{i=x}^{n} F_i}{\sum_{i=x}^{n} w_i} w_{px}$$

$$\geq 0.2 S_{DS} I_e w_{px}$$

$$\leq 0.4 S_{DS} I_e w_{px}$$

In the equation for F_{px}, w_i is the portion of the effective seismic weight, W, located at or assigned to level i and w_{px} is the weight tributary to the diaphragm at level x. For structures with walls, the weights of the walls parallel to the direction of analysis are often not included in w_{px} because these weights do not contribute to the diaphragm shear forces. However, including the weights of these walls in w_{px} is conservative, which means $w_{px} = w_x$ at level x.

The diaphragm forces are inertial forces and are applied at the CM at each level. The CM can occur at different locations on different levels, and this needs to be accounted for in the analysis.

The redundancy factor, ρ, is permitted to be set equal to 1.0 in the calculation of F_{px} by ASCE/SEI Equation (12.10-1) through (12.10-3) (ASCE/SEI 12.3.4.1).

In structures with a type 4 horizontal structural irregularity (out-of-plane offset irregularity; see Table 3.6), the transfer forces from the vertical seismic force–resisting elements above the diaphragm to the vertical seismic force–resisting elements below the diaphragm must be increased by the overstrength factor, Ω_0 (see Fig. 3.22). The modified transfer force is added to the diaphragm inertial force, F_{px}, originating on the diaphragm below.

A similar discontinuity is illustrated in Fig. 3.23 where reinforced concrete shear walls that are part of the SFRS above transfer forces through the reinforced concrete podium slab (diaphragm) to the reinforced concrete basement walls below. In addition to the inertial force of the podium slab, the diaphragm force includes the static soil pressure and dynamic seismic pressure, all of which get transferred to the basement walls parallel to the direction of analysis.

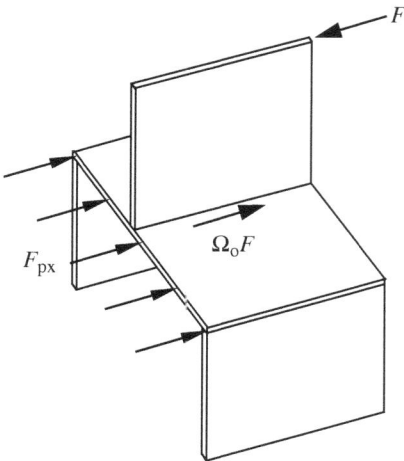

FIGURE 3.22 Seismic transfer force due to out-of-plane offset in the SFRS.

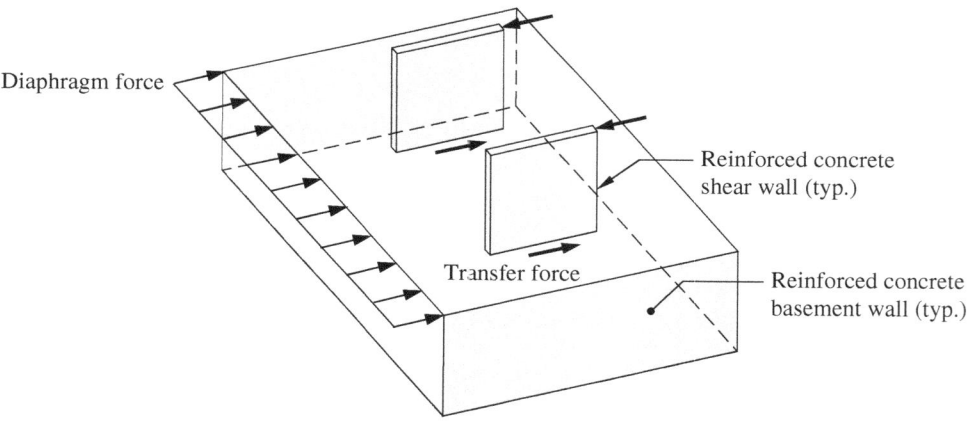

FIGURE 3.23 Transfer forces to a podium slab.

In cases with transfer forces, the redundancy factor, ρ, for the structure above must be applied to the transfer force, except where the transfer force must be increased by Ω_0.

For structures assigned to SDC D, E, or F with a type 1a, 1b, 2, 3, or 4 horizontal structural irregularity or a type 4 vertical structural irregularity, the diaphragm forces determined by ASCE/SEI 12.10.1.1 must be increased by 25 percent for the following elements of the SFRS (ASCE/SEI 12.3.3.4 and Table 3.8 of this publication):

- Connections of diaphragms to the vertical elements of the SFRS and to collectors.
- Collectors and their connections, including connections to vertical elements of the SFRS.

These forces need not be increased where the forces must be increased by Ω_0.

3.11.3 Collector Elements

Collectors must be designed for the combined effects due to gravity loads (flexure and shear) and lateral loads (axial tension and compression forces caused by in-plane shear transfer in the diaphragm due to seismic forces). The in-plane diaphragm forces are determined using the methods presented in Sec. 3.11.2 of this publication. Once the forces are determined, the diaphragm is modeled and analyzed using the methods in ASCE/SEI 12.3 (see Sec. 3.4 of this publication), and the axial forces in a collector can be determined based on the selected method of analysis.

According to ASCE/SEI 12.10.2.1, collectors and their connections to the vertical elements of the SFRS in buildings assigned to SDC C, D, E, or F must be designed to resist the effects from the maximum of the three forces given in Table 3.16.

The purpose of the overstrength requirements is to help ensure that inelastic behavior occurs in the ductile elements of the SFRS and not in the collectors or their connections. It is essential that the collectors and their connections perform as intended during a seismic event.

Collectors are to be designed using whichever of these three requirements produces the critical combined effects.

Where applicable, transfer forces must also be considered when determining forces in a collector (ASCE/SEI 12.10.2.1). As discussed in Sec. 3.11.2 of this publication, seismic transfer forces must be increased by the overstrength factor, Ω_o, in structures with a type 4 horizontal structural irregularity.

3.11.4 Alternative Design Provisions for Diaphragms, Including Chords and Collectors

Overview
As noted in Sec. 3.11.1 of this publication, the requirements in ASCE/SEI 12.10.3 must be used in the design of precast concrete diaphragms, including chords and collectors, in structures assigned to SDC C, D, E, or F and are permitted to be used for the design

Force Requirement	Load Combinations
Forces are calculated using the seismic load effects including overstrength of ASCE/SEI 12.4.3 with seismic forces determined by the ELF procedure of ASCE/SEI 12.8 or the modal response spectrum analysis procedure of ASCE/SEI 12.9.1.	In ASCE/SEI load combinations 6 and 7, use $E_{mh} = \Omega_o Q_E$ where Q_E is determined using F_x.
Forces are calculated using the seismic load effects including overstrength of ASCE/SEI 12.4.3 with seismic forces determined by ASCE/SEI Equation (12.10-1) for diaphragms.	In ASCE/SEI load combinations 6 and 7, use $E_{mh} = \Omega_o Q_E$ where Q_E is determined using F_{px}.
Forces are calculated using the load combinations of ASCE/SEI 2.3.6 with seismic forces determined by ASCE/SEI Equation (12.10-2), which is the lower-limit diaphragm force.	In ASCE/SEI load combinations 6 and 7, use $E_h = \rho Q_E$ where Q_E is determined using $F_{px} = 0.2 S_{DS} I_e w_{px}$.

TABLE 3.16 Collector Design Forces for Structures Assigned to SDC C, D, E, or F

of precast concrete diaphragms in structures assigned to SDC B, cast-in-place concrete diaphragms, and wood-sheathed diaphragms supported by wood diaphragm framing. Additionally, the following are applicable:

1. Footnote b in ASCE/SEI Table 12.2-1 pertaining to the permitted reduction of Ω_0 for structures with flexible diaphragms is not applicable.

2. ASCE/SEI 12.3.3.4 pertaining to the increase of forces caused by irregularities for SDC D through F is not applicable.

3. Item 5 in ASCE/SEI 12.3.4.1 pertaining to conditions where ρ is permitted to be taken as 1.0 is replaced with the following: "Design of diaphragms, including chords, collectors, and their connections to the vertical elements" are used.

4. Item 7 in ASCE/SEI 12.3.4.1 is not applicable.

Design
Diaphragms, including chords, collectors, and their connections to the vertical elements must be designed in two orthogonal directions to resist the in-plane design seismic force, F_{px}, determined in accordance with ASCE/SEI 12.10.3.2. Transfer of forces at diaphragm discontinuities, such as openings and reentrant corners, must also be considered in design.

Seismic Design Forces
The in-plane seismic design force, F_{px} is determined by ASCE/SEI Equation (12.10-4) and must be taken greater than or equal to the minimum force in ASCE/SEI Equation (12.10-5):

$$F_{px} = \frac{C_{px}}{R_s} w_{px} \geq 0.2 S_{DS} I_e w_{px} \tag{3.16}$$

The design acceleration coefficient, C_{px}, is determined by ASCE/SEI Figure 12.10-2 and depends on the number of stories in the building, N, the height of the floor level of interest from the base, h_x, and the structural height of the building, h_n. Design accelerations C_{p0}, C_{pi}, and C_{pn} determined by ASCE/SEI Equations (12.10-6), (12.10-8), (12.10-9), and (12.10-7), respectively, are needed in the determination of C_{px}. The following parameters are also required:

1. The modal contribution factors Γ_{m1} and Γ_{m2} determined by ASCE/SEI Equations (12.10-13) and (12.10-14), respectively.

2. The higher mode seismic response coefficient, C_{s2}, determined by ASCE/SEI Equations (12.10-10), (12.10-11), (12.10-12a), and (12.10-12b).

3. The mode shape factor, z_s, which is given as a function of the SFRS in ASCE/SEI 12.10.3.2.

Transfer Forces
In addition to the inertia forces, F_{px}, determined by ASCE/SEI Equations (12.10-4) and (12.10-5), all diaphragms must be designed for all applicable transfer forces (ASCE/SEI 12.10.3.3). As noted in Sec. 3.11.2 of this publication, transfer forces from the vertical seismic force–resisting elements above the diaphragm to the vertical

seismic force–resisting elements below the diaphragm must be increased by the overstrength factor, Ω_0, in structures with a type 4 horizontal structural irregularity (see Fig. 3.22). In the case of one- and two-family dwellings of light-frame construction, it is permitted to use $\Omega_0 = 1.0$ (see the exception in ASCE/SEI 12.10.3.3).

Collectors—SDCs C through F

According to ASCE/SEI 12.10.3.4, collectors and their connections, including connections to vertical elements of the SFRS, in buildings assigned to SDC C, D, E, or F must be designed for the effects due to 1.5 times F_{px} determined in accordance with ASCE/SEI 12.10.3.2. Where transfer forces are applicable, the effects from the transfer forces must also be multiplied by 1.5.

Three exceptions to these requirements are given in ASCE/SEI 12.10.3.4.

Diaphragm Design Force Reduction Factor, R_s

Values of R_s for various diaphragm systems are given in ASCE/SEI Table 12.10-1. Explanation and derivation of these factors are given in Ref. 3.

The flowchart in Fig. 3.24 can be used to determine the design seismic diaphragm force, F_{px}, in accordance with ASCE/SEI 12.10.3.

3.12 Structural Walls and Their Anchorage

3.12.1 Design for Out-of-Plane Forces

In addition to forces in the plane of the wall, structural walls and their anchorage in structures assigned to SDC B and higher must be designed for an out-of-plane force F_p equal to $0.4S_{DS}I_e$ times the weight of the structural wall or 0.10 times the weight of the structural wall, whichever is greater (ASCE/SEI 12.11.1). This force acts perpendicular to the face of the wall.

Because walls are often subjected to local deformations due to material shrinkage, temperature changes, and foundation settlement, structural wall elements and connections to supporting framing systems require some degree of ductility in order to accommodate these deformations while providing the required strength for combined gravity and seismic forces.

Nonstructural walls need not be designed for this requirement but must be designed in accordance with the seismic design for nonstructural component requirements in ASCE/SEI Chapter 13.

3.12.2 Anchorage of Structural Walls

Structural walls in structures assigned to SDC B and higher must be adequately anchored to the roof and floor members that provide lateral support for the walls. The anchorage of structural walls must provide a direct connection that is capable of resisting the force F_p determined by ASCE/SEI Equation (12.11-1):

$$F_p = 0.4S_{DS}k_aI_eW_p \geq 0.2k_aI_eW_p \qquad (3.17)$$

where W_p is the weight of the wall tributary to the anchor.

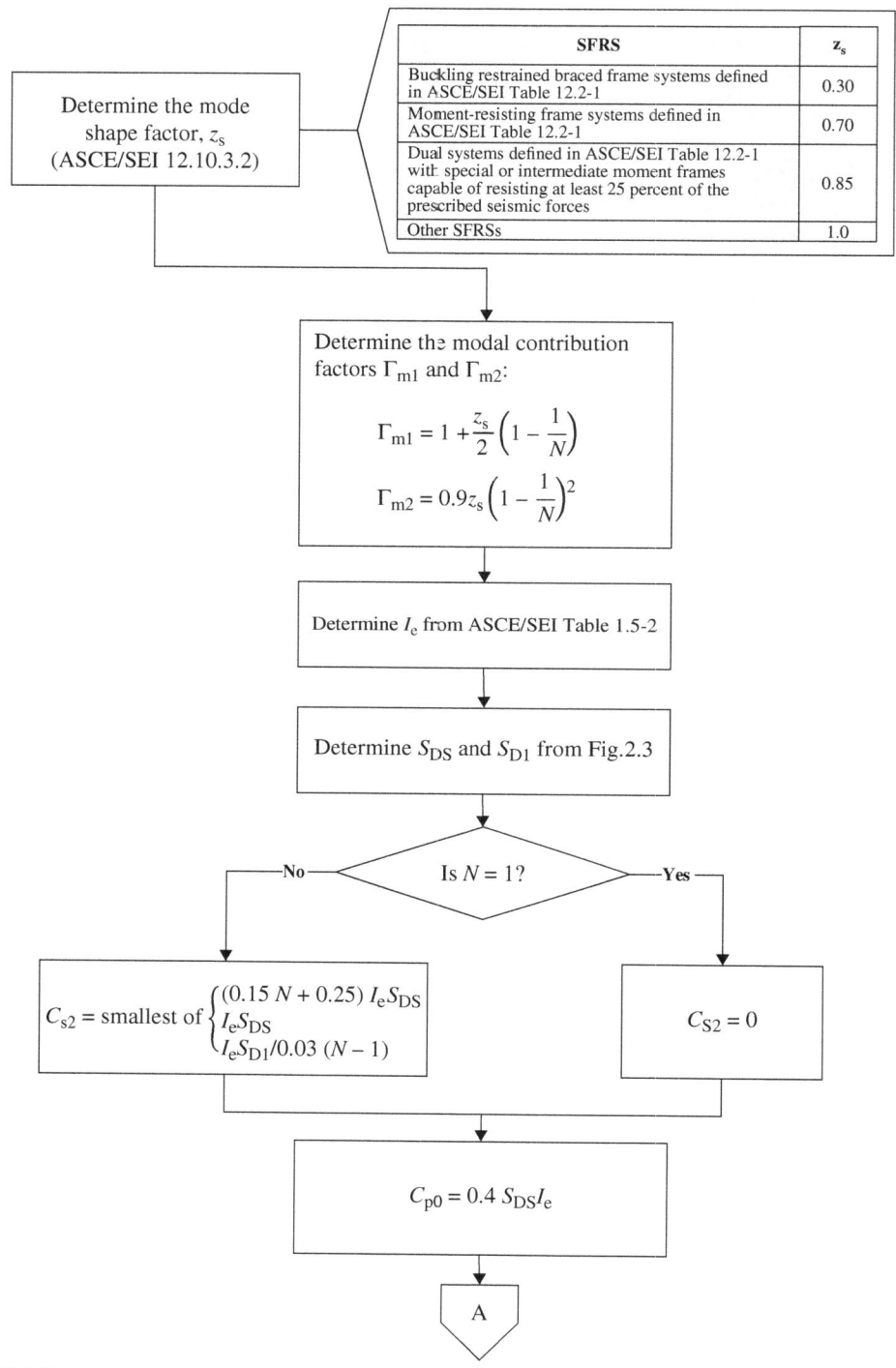

SFRS	z_s
Buckling restrained braced frame systems defined in ASCE/SEI Table 12.2-1	0.30
Moment-resisting frame systems defined in ASCE/SEI Table 12.2-1	0.70
Dual systems defined in ASCE/SEI Table 12.2-1 with special or intermediate moment frames capable of resisting at least 25 percent of the prescribed seismic forces	0.85
Other SFRSs	1.0

Determine the mode shape factor, z_s (ASCE/SEI 12.10.3.2)

Determine the modal contribution factors Γ_{m1} and Γ_{m2}:

$$\Gamma_{m1} = 1 + \frac{z_s}{2}\left(1 - \frac{1}{N}\right)$$

$$\Gamma_{m2} = 0.9z_s\left(1 - \frac{1}{N}\right)^2$$

Determine I_e from ASCE/SEI Table 1.5-2

Determine S_{DS} and S_{D1} from Fig.2.3

Is $N = 1$?

No

Yes

$$C_{s2} = \text{smallest of} \begin{cases} (0.15\,N + 0.25)\,I_e S_{DS} \\ I_e S_{DS} \\ I_e S_{D1}/0.03\,(N-1) \end{cases}$$

$$C_{S2} = 0$$

$$C_{p0} = 0.4\,S_{DS}I_e$$

A

FIGURE 3.24 Design seismic diaphragm force in accordance with ASCE/SEI 12.10.3.

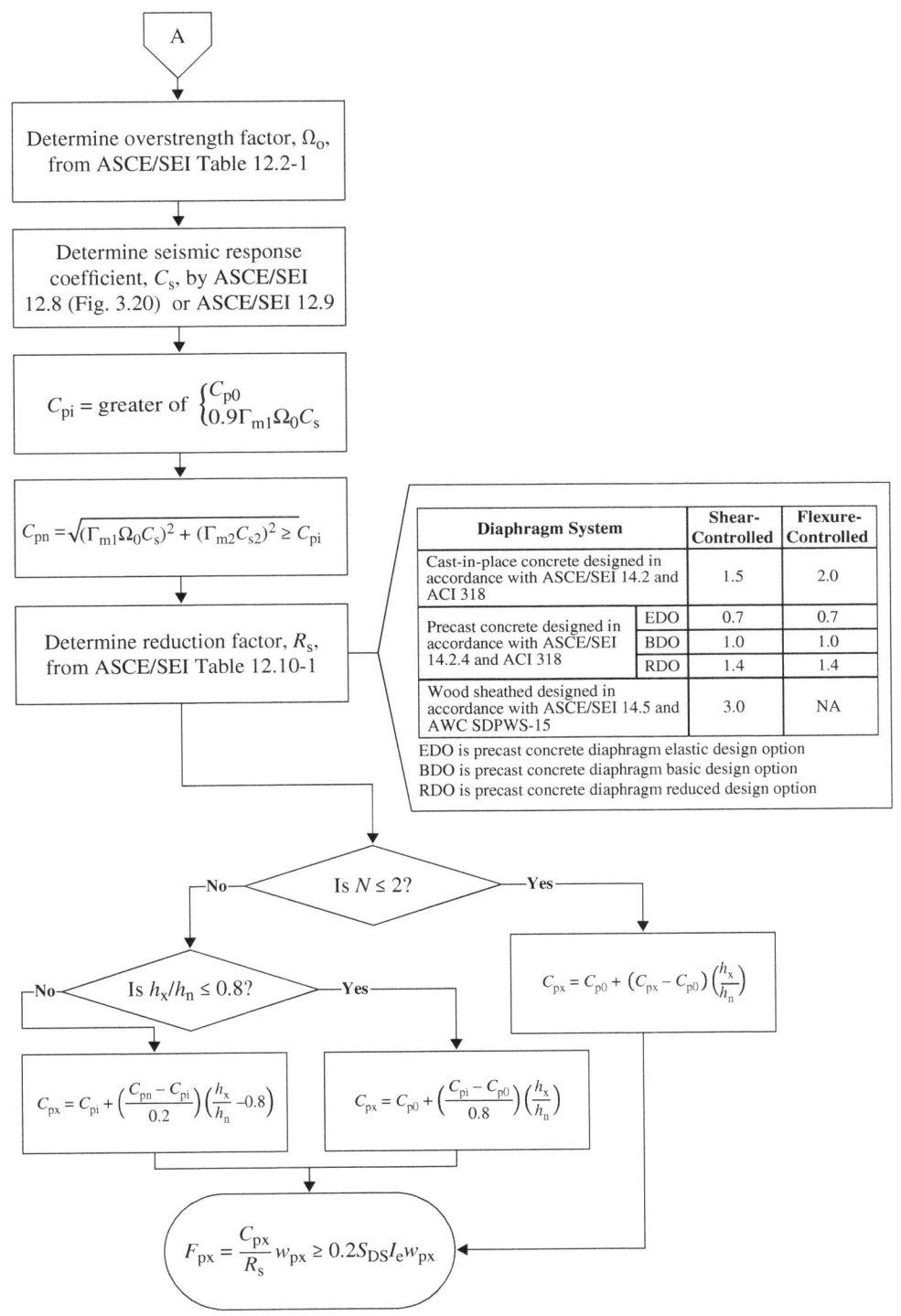

FIGURE 3.24 (Continued)

The amplification factor for diaphragm flexibility, k_a, which accounts for amplification of out-of-plane accelerations caused by diaphragm flexibility, is determined by ASCE/SEI Equation (12.11-2):

$$k_a = \begin{cases} 1.0 + \dfrac{L_f}{100} \leq 2.0 \text{ for flexible diaphragms} \\ 1.0 \text{ for diaphragms that are not flexible} \end{cases} \tag{3.18}$$

where L_f is the span, in feet, of a flexible diaphragm that provides support for the wall; the span is measured between vertical elements that provide lateral support to the diaphragm in the direction of analysis ($L_f = 0$ for rigid diaphragms).

The purpose of this requirement is to help prevent separation of structural walls from the roof and floors. The force F_p applies only to the design of the anchorage or connection of the wall to the structure and not to the overall design of the wall. This force is to be applied for both for tension (out-of-plane) and sliding (in-plane) where applicable (for example, where roof or floor framing is not perpendicular to the anchored walls).

In a structure where all the diaphragms are not flexible, the value of F_p determined by Eq. (3.17) is permitted to be multiplied by the factor $(1 + 2z/h)/3$ where anchorage is not located at the roof. The height of the anchor above the base of the structure is designated z and h is the height of the roof above the base. This reduction factor can be conservatively set equal to 1.0 to ensure that a smaller than appropriate anchorage force is not used in design. In this case, F_p must be greater than or equal to that required by ASCE/SEI 12.11.2 with a minimum anchorage force of $0.2W_p$.

Where the anchor spacing is greater than 4 ft (1.2 m) along the length of the wall, the section of wall spanning between the anchors must be designed to resist the local out-of-plane bending caused by F_p.

3.12.3 Additional Requirements for Anchorage of Concrete and Masonry Structural Walls to Diaphragms in Structures Assigned to SDC C through F

Additional requirements for diaphragms in structures assigned to SDC C through F are given in ASCE/SEI 12.11.2.2. The main purpose of these requirements is to ensure that a continuous load path exists that will survive the ground shaking.

3.13 Drift and Deformation

3.13.1 Overview

Story drift must be controlled for both structural and nonstructural reasons. Limiting the drift helps limit inelastic strain in structural members and helps control overall structural stability by reducing P-delta effects. Drift control is also needed to restrict damage to nonstructural elements such as partitions, elevator and stair enclosures, and glass, to name a few.

3.13.2   Story Drift Limit

Design drifts are determined in accordance with ASCE/SEI 12.8.6 (see Sec. 3.9.7 of this publication); these drifts must be less than or equal to the allowable story drift, Δ_a, in ASCE/SEI Table 12.12-1. The drift limits depend on the risk category: drift limits are generally more restrictive for Risk Categories III and IV and are meant to provide a higher level of performance for more important structures.

Drift limits also depend on the type of structure. For ordinary structures, the drift limit is 2 percent of the story height. For low-rise structures (four stories or less) where the interior walls, partitions, ceilings, and exterior wall systems have been designed to accommodate story drifts, the drift limits are not as stringent as those for other types of structures.

Satisfying strength requirements may result in a structure that also satisfies story drift limits. Moment-resisting frames or tall, slender structures are often controlled by drift and member sizes of the SFRS generally have to be increased to satisfy prescribed drift limits.

3.13.3   Moment Frames in Structures Assigned to SDC D through F

For structures assigned to SDC D, E or F with SFRSs consisting of only moment-resisting frames, the design story drift, Δ, must be less than or equal to Δ_a/ρ for any story (ASCE/SEI 12.12.1.1). Given the inherent flexibility of moment-resisting systems, this provision essentially penalizes structures utilizing nonredundant systems in regions of high seismic risk.

3.13.4   Diaphragm Deflection

Elements attached to diaphragms must be able to maintain their structural integrity and to support any applicable loads as the diaphragm deflects in its own plane due to the seismic forces (ASCE/SEI 12.12.2). The permissible in-plane deflection of a diaphragm is equal to the permissible deflections of the attached elements. Diaphragm deflections can be determined by a variety of methods based on the flexibility of the diaphragm.

3.13.5   Structural Separation

Portions of a structure or adjacent structures must have sufficient distance between them so that they can respond independently to ground motion without impacting.

The maximum inelastic response displacement δ_M occurring at the critical location is used in determining separation distances (see ASCE/SEI Equation (12.12-1)):

$$\delta_M = \frac{C_d \delta_{max}}{I_e} \tag{3.19}$$

In this equation, δ_{max} is the maximum elastic displacement at the critical location, which is determined based on the seismic story forces F_x and includes translational and torsional displacements of the structure. Where applicable, amplification of the accidental torsional deflections as prescribed in ASCE/SEI 12.8.4.3 must also be included in the determination of δ_{max}.

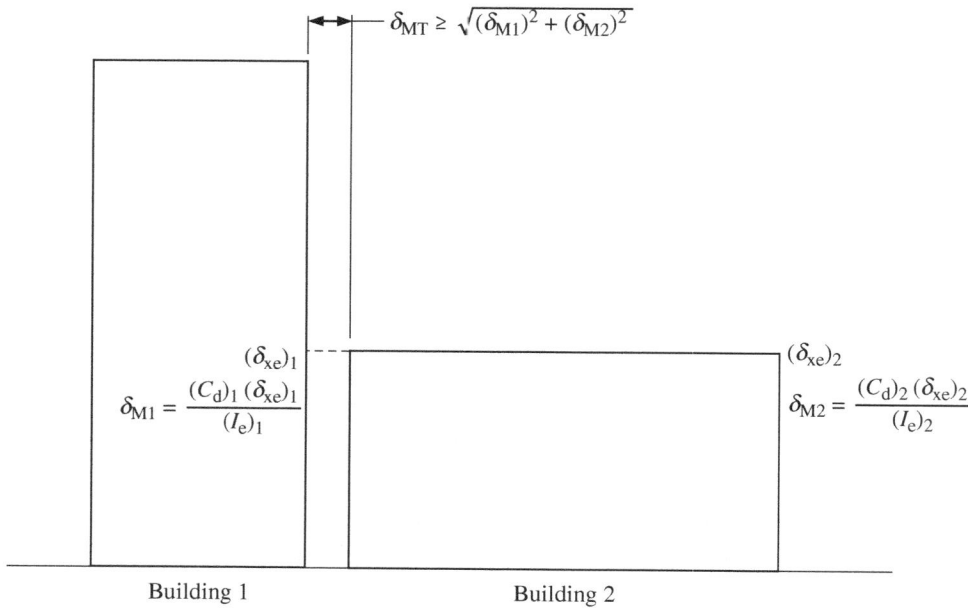

FIGURE 3.25   Structural separation requirements in accordance with ASCE/SEI 12.12.3.

Once δ_M has been calculated for each of the adjacent structures, the minimum separation distance, δ_{MT}, is determined by ASCE/SEI Equation (12.12-2):

$$\delta_{MT} = \sqrt{(\delta_{M1})^2 + (\delta_{M2})^2}$$ (3.20)

where δ_{M1} and δ_{M2} are the maximum inelastic response displacements of the adjacent structures at their adjacent edges (see Fig. 3.25).

The above requirements are applicable to single buildings with one or more internal separation joints (commonly referred to seismic joints) or to adjacent buildings on the same property where sufficient information is known about both structures so that the required displacements can be calculated.

Where a new building will be located in proximity to an existing structure, it is unlikely information about the existing structure is available. In such cases, separation calculations must be performed based on the maximum deflections of the new building. For example, if the taller building in Fig. 3.25 is the new building, the new building must be set back a distance of $\delta_M = C_d \delta_{roof} / I_e$ from the existing building.

3.13.6   Members Spanning between Structures

For members connected to adjacent structures or to seismically separated portions of structures, the gravity connections or supports of such members must be designed to accommodate the maximum anticipated relative displacements. These displacements are to be calculated based on the following (ASCE/SEI 12.12.4):

1. The deflections at the locations of support are to be calculated using ASCE/SEI Equation (12.8-15) multiplied by $1.5R/C_d$.

2. Additional deflection due to diaphragm rotation (including the torsional amplification factor of ASCE/SEI 12.8.4.3, where applicable) must be included where either structure is torsionally irregular.

3. Diaphragm deformations must be included.

4. The two structures are assumed to be moving in opposite directions and the absolute sum of the displacements must be used.

3.13.7 Deformation Compatibility

Certain members or components in a structure are assigned to be part of the SFRS, which means these elements must be designed for the combined effects due to gravity and seismic loads. During an earthquake, all the structural members undergo displacement due to the ground motion regardless if they are part of the SFRS or not.

For structures assigned to SDC D through F, structural members not included in the SFRS must be designed for deformation compatibility: these members must be designed to adequately resist gravity load effects when subjected to the design story drift Δ determined in accordance with ASCE/SEI 12.8.6 (ASCE/SEI 12.12.5). The story drifts induce seismic load effects into these members that must be accounted for in design.

The stiffening effects of adjoining rigid structural and nonstructural elements must be considered in the design of the members not included in the SFRS.

3.14 Foundation Design

General requirements for the design of various types of foundation systems are given in ASCE/SEI 12.13 and Chapter 18 of the IBC. The special seismic requirements are based on the SDC.

Both strength design and allowable stress design of foundation elements are permitted (see ASCE/SEI 12.13.5 and 12.13.6, respectively). For strength design, the nominal foundation geotechnical capacity, Q_{ns}, is determined by one of the following methods (ASCE/SEI 12.13.5.1):

- Presumptive load-bearing values.
- By a registered design professional based on geotechnical site investigations that include field and laboratory testing.
- By in situ testing of prototype foundations.

Once Q_{ns} has been determined, the design strength is determined by multiplying Q_{ns} by the resistance factors, ϕ, for strength design in ASCE/SEI Table 12.13-1. Resistance factors are provided for vertical resistance (bearing strength and pile friction) and lateral resistance (lateral bearing pressure and sliding by friction or cohesion). The required strength is obtained by using the strength design load combinations in ASCE/SEI 2.3; these factored load effects, including reductions permitted in ASCE/SEI 12.13.4, must be less than or equal to the corresponding design strength, ϕQ_{ns}.

In the allowable stress method, the factored load effects determined by the allowable stress design load combinations in ASCE/SEI 2.4 must be less than or equal to the allowable foundation load capacities, Q_{as}, which are determined by geotechnical investigations required by the authority having jurisdiction. The reduction of foundation overturning effects in ASCE/SEI 12.13.4 is permitted to be used.

Provisions for foundations on liquefiable sites are given in ASCE/SEI 12.13.9.

3.15 Simplified Alternative Structural Design Criteria for Simple Bearing Wall or Building Frame Systems

3.15.1 Overview

The simplified method in ASCE/SEI 12.14 is permitted to be used in lieu of other analytical procedures in ASCE/SEI Chapter 12 for the analysis and design of simple bearing wall or building frame systems provided the twelve limitations of ASCE/SEI 12.14.1.1 are satisfied (see Table 3.17). Even though twelve conditions must be met, the procedure is applicable to a wide range of relatively stiff, low-rise structures that fall under risk categories I and II and possess SFRSs arranged in a torsionally resistant, regular layout.

3.15.2 Seismic Load Effects and Combinations

The seismic load effect, E, in the simplified method is defined the same as in the general requirements of ASCE 12.4 and consists of the effects of horizontal and vertical

Number	Limitation
1	Risk Category I or II structures
2	Site Class A through D
3	Structure must not exceed three stories above grade.
4	SFRS must be bearing wall or building frame as indicated in ASCE/SEI Table 12.14-1.
5	Two lines of walls or frames are required in each of two major axis directions where at least one line of resistance must be provided on each side of the CM in each direction.
6	The CM in each story must be located not further from the geometric centroid of the diaphragm than 10 percent of the length of the diaphragm parallel to the eccentricity.
7	ASCE/SEI Equation (12.14-1) must be satisfied for structures with cast-in-place concrete diaphragms with overhangs beyond the outside line of shear walls or braced frames.
8	For structures with diaphragms that are not flexible, the forces are to be apportioned to the vertical elements of the SFRS as if the diaphragms were flexible and the requirements in 8a and 8b must be satisfied.
9	Lines of walls or frames must be oriented at angles of no more than 15 degrees from the major orthogonal horizontal axes of the building.
10	The simplified procedure must be used for each major orthogonal horizontal axis of the building.
11	System irregularities caused by in-plane or out-of-plane offsets of lateral force–resisting elements are not permitted.
12	Lateral load resistance of any story must be at least 80 percent of the story above.

TABLE 3.17 Limitations of the Simplified Seismic Procedure of ASCE/SEI 12.14

earthquake-induced forces E_h and E_v, respectively (see ASCE/SEI 12.14.3.1 and Sec. 3.5 of this publication). In the simplified method, it is assumed the structure possesses a reasonable level of redundancy (see the fifth limitation in Table 3.17). Therefore, the redundancy factor, ρ, is equal to 1.0 in the simplified method, and E_h is equal to Q_E.

The vertical seismic load effect, E_v, is permitted to be taken as zero for either of the following conditions:

1. In ASCE/SEI Equations (12.14-3), (12.14-4), (12.14-7), and (12.14-8) where $S_{DS} \leq 0.125$.
2. In ASCE/SEI Equation (12.14-4) where determining demands on the soil-structure interface of foundations.

The seismic load combinations in ASCE/SEI 12.14 are the same as those in ASCE/SEI 12.4. In the simplified method, $\Omega_0 = 2.5$, which is consistent with the values of Ω_0 in ASCE/SEI Table 12.2-1 for bearing wall and building frame systems.

3.15.3 Seismic Force–Resisting Systems

Selection and Limitations
Parts of ASCE/SEI Table 12.2-1 are reproduced in ASCE/SEI Table 12.14-1 for the systems permitted to be designed by the simplified method. The sections pertaining to detailing requirements, the response modification coefficient, R, and system limitations for SDCs B through E are given in ASCE/SEI Table 12.14-1. As noted above, the system over-strength factor, Ω_0, is taken as 2.5 for these systems, so it is not included in the table. Also, because drift calculations are not required (see ASCE/SEI 12.14.8.5 and the discussion below), the deflection amplification factor, C_d, is not in ASCE/SEI Table 12.14-1.

Combinations of Framing Systems
Combinations of framing systems are permitted horizontally and vertically (see ASCE/SEI 12.14.4.2). The value of R used in design must be the least value of any of the SFRSs in the direction of analysis.

3.15.4 Diaphragm Flexibility

Untopped metal deck, wood structural panels, and similar panelized constructions are permitted to be considered flexible diaphragms (ASCE/SEI 12.14.5), and lateral load is distributed to the vertical elements of the SFRS using tributary area (ASCE/SEI 12.14.8.3.1).

For diaphragms that are not flexible, a rigidity analysis is required, which includes torsional moments resulting from eccentricity between the locations of CM and CR (ASCE/SEI 12.14.8.3.2). Accidental torsion and dynamic amplification of torsion need not be included in the analysis because the simplified method is applicable to regular structures with essentially a uniform distribution of lateral stiffness.

3.15.5 Application of Loading

Design seismic forces, which are determined in accordance with ASCE/SEI 12.14.8 (see discussion in the following text), are permitted to be applied separately in each of the orthogonal directions (ASCE/SEI 12.14.6). In other words, two separate analyses are acceptable, and the combination of load effects from the two directions need not be considered.

3.15.6 Design and Detailing Requirements

A summary of the design and detailing requirements in ASCE/SEI 12.14.7 is given in Table 3.18.

3.15.7 Simplified Lateral Force Analysis Procedure

A summary of the requirements in ASCE/SEI 12.14.8 for the simplified lateral force analysis procedure is given in Table 3.19.

The flowchart in Fig. 3.26 can be used to determine the design seismic forces and their distribution based on the requirements of the simplified method in ASCE/SEI 12.14.

3.16 Examples

The following examples illustrate the determination of seismic forces and their effects on building structures based on the requirements given in this chapter.

3.16.1 Example 3.1—Combination of Framing Systems in the Same Direction: Vertical Combinations

Determine the governing design coefficients R, Ω_0, and C_d in the north-south direction for the commercial building in Fig. 3.27. The SFRS consists of an ordinary reinforced concrete shear wall (nonbearing) supported by an ordinary reinforced concrete moment frame. Assume the building is assigned to SDC B.

Solution

The requirements in ASCE/SEI 12.2.3 are used to determine the governing design coefficients.

Because the upper rigid portion is supported by a flexible lower portion, a two-stage analysis procedure is not permitted (ASCE/SEI 12.2.3.2). Therefore, use the requirements in ASCE/SEI 12.2.3.1.

Step 1—Determine R, Ω_0, and C_d for each system　　　　　ASCE/SEI Table 12.2-1

For the ordinary reinforced concrete shear walls (system B5): $R = 5$, $\Omega_0 = 2.5$, and $C_d = 4.5$.

For the ordinary reinforced concrete moment frames (system C7): $R = 3$, $\Omega_0 = 3$, and $C_d = 2.5$.

Step 2—Determine the governing design coefficients　　　　　ASCE/SEI 12.2.3.1

The design coefficients for the upper system are permitted to be used to calculate the forces and drifts on the upper system because R of the lower system is less than R of the upper system (see Fig. 3.3). For the design of the lower system, the design coefficients of the lower system must be used.

The force transferred from the shear wall to the moment frame must be increased by the ratio R of the upper system to R of the lower system, that is, by the factor $5/3 = 1.67$.

TABLE 3.18 Design and Detailing Requirements in ASCE/SEI 12.14.7

ASCE/SEI Section Number	Requirement	Exceptions/Remarks
12.14.7.1 Connections	All parts of the structure must be adequately interconnected to achieve a continuous load path through a structure down to the foundation.	Design and detailing requirements are independent of the SDC. The requirements for interconnection in ASCE/SEI 12.14.7.1 are somewhat more stringent than those in ASCE/SEI 12.1.3, while the requirements for connections to supports are the same in ASCE/SEI 12.1.4 and 12.14.7.1. Foundation design requirements are the same in ASCE/SEI 12.14.7 and 12.1.5.
12.14.7.2 Openings or reentrant building corners	Reinforcement must be provided at the edges of openings in shear walls, diaphragms, or other plate-type elements and at reentrant corners.	Shear walls of wood structural panels are permitted where designed in accordance with AWC SDPWS-15 for perforated shear walls or ANSI/AISI S400 for type II shear walls.
12.14.7.3 Collector elements	Collector elements and their connections must be designed to resist the seismic load effects including an overstrength factor equal to 2.5.	In structures that are braced entirely by light-frame shear walls, it is permitted to design collectors and their connections to resist the forces prescribed in ASCE/SEI 12.14.7.4.
12.14.7.4 Diaphragms	Diaphragms must be designed to resist the design seismic force, F_x, determined by ASCE/SEI Equation (12.14-13). Where applicable, transfer forces from the SFRS above must be added to the diaphragm design force.	—
12.14.7.5 Anchorage of structural walls	Structural walls must be anchored to floors, roofs, and members providing out-of-plane lateral support for the wall. The anchorage must resist the force determined by ASCE/SEI Equation (12.14-10).	The equation to determine the design force in individual anchors is the same as that in ASCE/SEI 12.11.2.1 (see Sec. 3.12 of this publication).
12.14.7.6 Bearing walls and shear walls	Exterior and interior bearing walls and shear walls and their anchorage must be designed to resist a force normal to the surface equal to $0.4S_{DS}W_p \geq 0.1W_p$.	The requirements for out-of-plane forces on exterior and interior structural walls are the same as those in ASCE/SEI 12.11.1.
12.14.7.7 Anchorage of nonstructural components	All portions or components of the structure must be anchored to resist the seismic force F_p determined in accordance with ASCE/SEI Chapter 13, where required.	—

113

ASCE/SEI Section Number	Requirement	Exceptions/Remarks
12.14.8	Modeling	A linear mathematical model of the structure must be constructed. For purposes of analysis, the structure is permitted to be fixed at the base.	—
12.14.8.1	Seismic base shear	Seismic base shear, V, is determined by ASCE/SEI Equation (12.14-2): $$V = \frac{F S_{DS}}{R} W$$ where $$F = \begin{cases} 1.0 \text{ for one-story buildings} \\ 1.1 \text{ for two-story buildings} \\ 1.2 \text{ for three-story buildings} \end{cases}$$ $$S_{DS} = 2F_a S_S / 3$$ In calculating S_{DS}, S_S is determined in accordance with ASCE/SEI 11.4.4 but need not be taken larger than 1.5. $$F_a = \begin{cases} 1.0 \text{ for rock sites} \\ 1.4 \text{ for soil sites} \end{cases}$$ Sites are permitted to be considered rock where there is no more than 10 ft (3.0 m) of soil between the rock surface and the bottom of the foundation. W = effective seismic weight defined in ASCE/SEI 12.14.8.1 R = response modification coefficient given in ASCE/SEI Table 12.14-1	It is evident that V, which is independent of the building period, is increased by 10 and 20 percent for two-story and three-story buildings, respectively. These increases primarily account for the method used for vertical distribution of V, which is based on tributary weight, and have been shown by parametric studies to be adequate without being overly conservative. In lieu of determining the short-period site coefficient, F_a, in accordance with ASCE/SEI 11.4.3, values of F_a are permitted to be taken as given in this section. Only a basic geotechnical investigation is needed; 100-ft (30.5-m) deep borings and seismic shear velocity tests are not necessary.
12.14.8.2	Vertical distribution	Seismic forces at each level are determined by ASCE/SEI Equation (12.14-13): $$F_x = \frac{w_x}{W} V$$ where w_x is the portion of W at level x	The vertical distribution of the seismic forces is triangular.

114

12.14.8.3	Horizontal shear distribution	The seismic design story shear is determined by ASCE/SEI Equation (12.14-14): $$V_x = \sum_{i=x}^{n} F_i$$ where F_i is the portion of V induced at level i. For structures with flexible diaphragms, the seismic design story shear is distributed to the vertical elements of the SFRS based on the area tributary to the vertical element. For structures with diaphragms that are not flexible, the seismic design story shear is distributed to the vertical elements of the SFRS based on the relative lateral stiffnesses of the vertical elements and the diaphragm. The inherent torsional moment, M_T, resulting from the eccentricity between the locations of the CM and CR must be included in the analysis.	These requirements are the same as those for the ELF procedure. A two-dimensional analysis is permitted for structures with flexible diaphragms.
114.8.4	Overturning	The structure must be designed to resist the effects from overturning caused by the seismic forces. Foundations of structure must be designed for not less than 75 percent of the foundation overturning moment, M_t, at the foundation-soil interface.	These requirements are the same as those for the ELF procedure. Included is the 25 percent reduction that can be applied when considering foundation overturning.
12.14.8.5	Drift limits and building separation	Structural drift need not be calculated. Where drift calculations are needed to determine structural separation between buildings or from property lines, for design of cladding, or for other design requirements, the drift must be taken as 1 percent of the structural height, h_n, unless computed to be less.	The simplified method does not require a drift check because it is assumed that bearing wall and building frame systems within the prescribed height range will not require such a check (unlike moment frame systems where drift is a major concern in design). Accordingly, calculations for P-delta effects need not be considered.

TABLE 3.19 Simplified Lateral Force Analysis Procedure

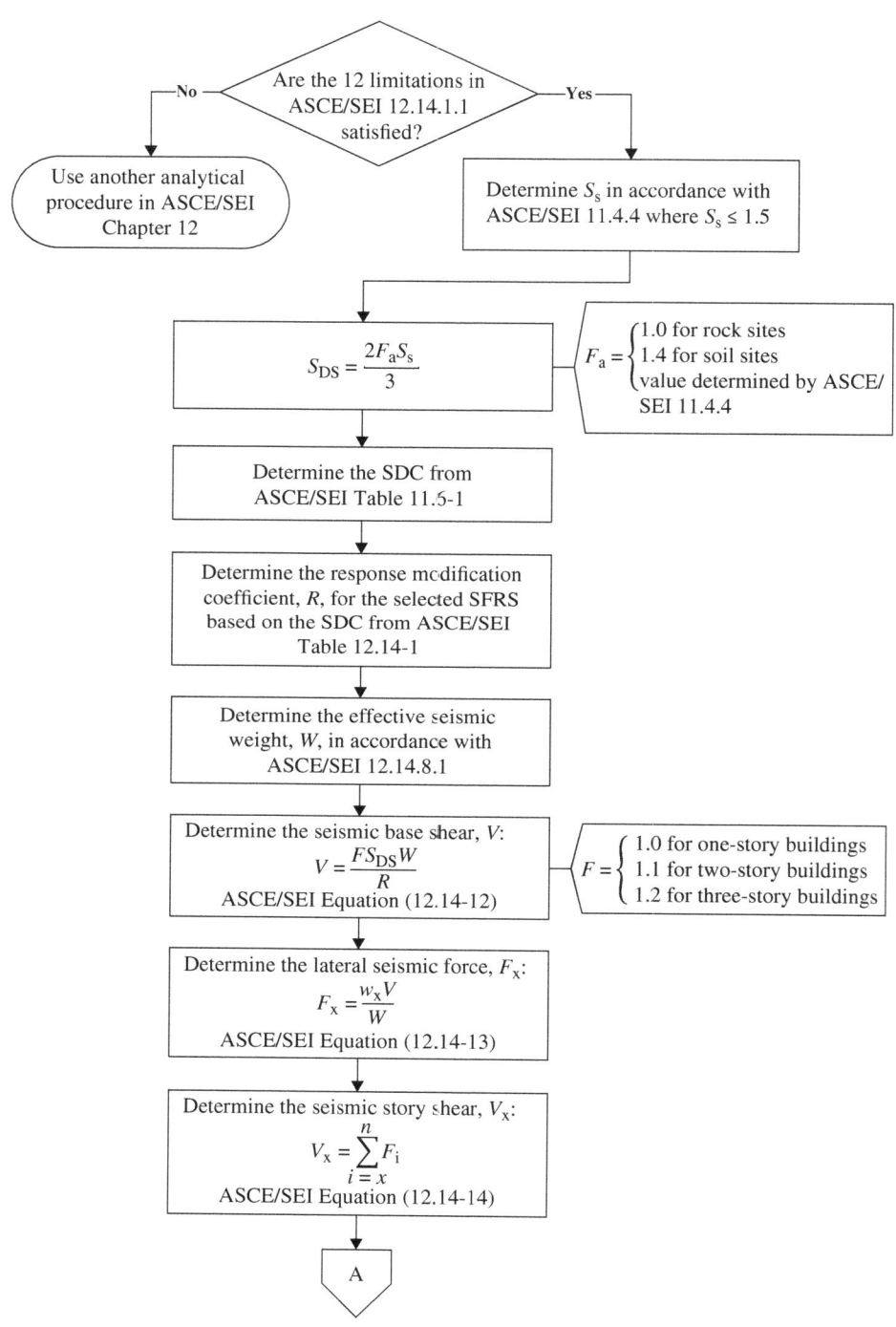

FIGURE 3.26 Design seismic forces in accordance with the simplified alternative structural design criteria in ASCE/SEI 12.14.

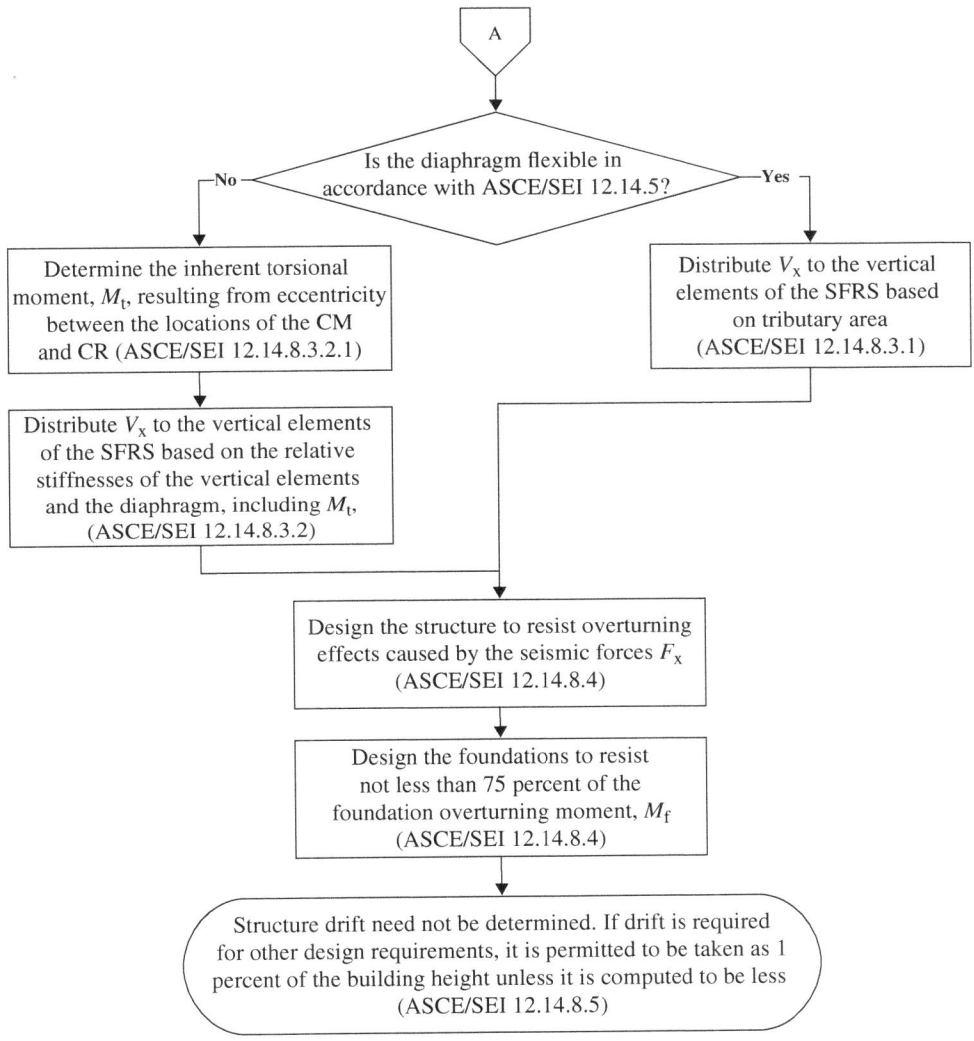

A

Is the diaphragm flexible in accordance with ASCE/SEI 12.14.5?

No — Yes

Determine the inherent torsional moment, M_t, resulting from eccentricity between the locations of the CM and CR (ASCE/SEI 12.14.8.3.2.1)

Distribute V_x to the vertical elements of the SFRS based on tributary area (ASCE/SEI 12.14.8.3.1)

Distribute V_x to the vertical elements of the SFRS based on the relative stiffnesses of the vertical elements and the diaphragm, including M_t, (ASCE/SEI 12.14.8.3.2)

Design the structure to resist overturning effects caused by the seismic forces F_x (ASCE/SEI 12.14.8.4)

Design the foundations to resist not less than 75 percent of the foundation overturning moment, M_f (ASCE/SEI 12.14.8.4)

Structure drift need not be determined. If drift is required for other design requirements, it is permitted to be taken as 1 percent of the building height unless it is computed to be less (ASCE/SEI 12.14.8.5)

FIGURE 3.26 *(Continued)*

3.16.2 Example 3.2—Combination of Framing Systems in the Same Direction: Vertical Combinations

Determine the governing design coefficients R, Ω_0, and C_d in the north-south direction for the commercial building in Example 3.1 assuming the ordinary reinforced concrete moment frame is supported by the ordinary reinforced concrete shear wall, which is 8 in. (203 mm) thick (see Fig. 3.28).

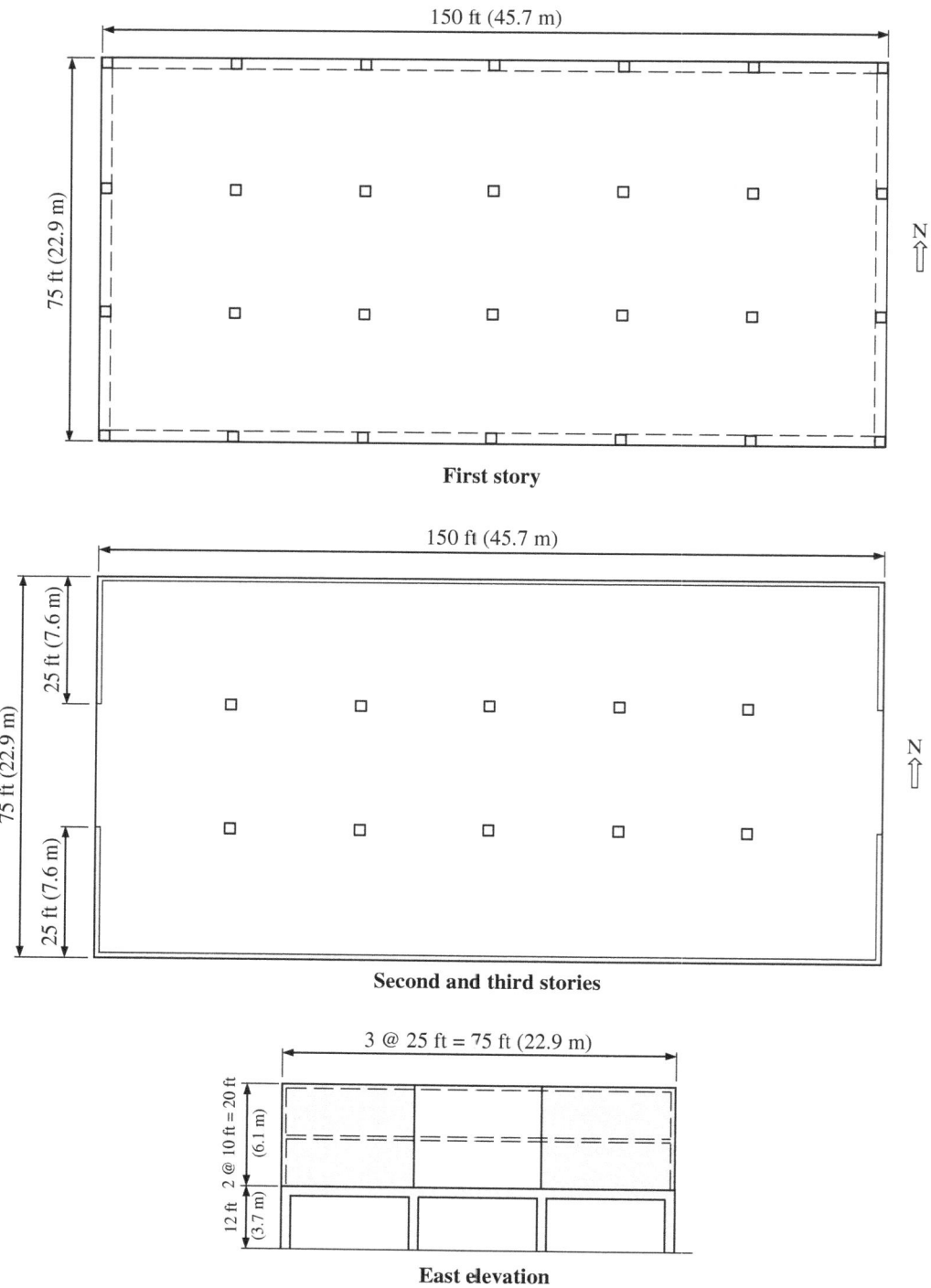

First story

Second and third stories

East elevation

FIGURE 3.27   Vertical combination of framing systems for the building in Example 3.1.

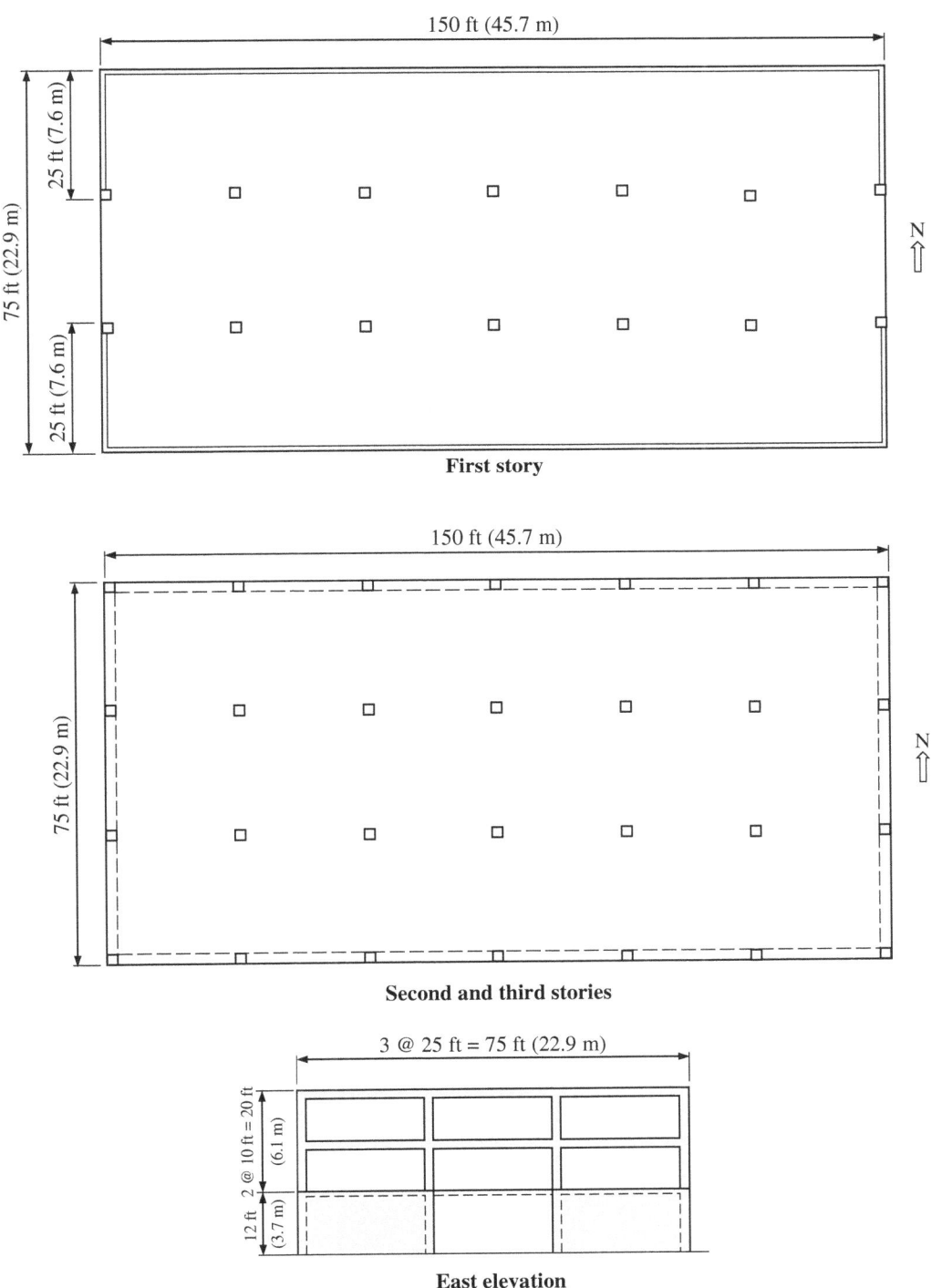

FIGURE 3.28 Vertical combination of framing systems for the building in Example 3.2.

Solution

The requirements in ASCE/SEI 12.2.3 are used to determine the governing design coefficients.

Step 1—Determine if a two-stage analysis procedure can be used

ASCE/SEI 12.2.3.2

The vertical combination of SFRSs in this building consist of a flexible upper portion supported by a rigid lower portion. A two-stage equivalent lateral force procedure is permitted to be used provided the five conditions in ASCE/SEI 12.2.3.2 are satisfied:

a. The stiffness of the lower portion must be at least 10 times the stiffness of the upper portion.

It can be determined that the lateral stiffness of the lower portion is more than 10 times the lateral stiffness of the upper portion.

b. The period of the entire structure must be less than or equal to 1.1 times the period of the upper portion considered as a separate structure supported at the transition from the upper to the lower portion.

The approximate period of the moment frame in the upper portion is determined using ASCE/SEI Equation (12.8-7) and the approximate period parameters in ASCE/SEI Table 12.8-2:

For concrete moment-resisting frames, $C_t = 0.016$ (in S.I.: 0.0466) and $x = 0.9$.

$$T_a = C_t h_n^x = 0.016 \times 20^{0.9} = 0.24 \text{ s}$$

In S.I.: $T_a = 0.0466 \times 6.1^{0.9} = 0.24$ s

For the concrete shear walls (ASCE/SEI Equation (12.8-9)):

$$T_a = \frac{C_q}{\sqrt{C_w}} h_r$$

$$C_q = 0.0019 \text{ ft } (0.0058 \text{ m})$$

$$C_w = \frac{100}{A_B} \sum_{i=1}^{x} \frac{A_i}{\left[1 + 0.83\left(\dfrac{h_n}{D_i}\right)^2\right]}$$

A_B = area of base of structure = $75 \times 150 = 11{,}250 \text{ ft}^2$

A_i = web area of shear wall $i = (8/12) \times 25 = 16.7 \text{ ft}^2$

D_i = length of shear wall $i = 25$ ft

$$C_w = \frac{100}{11{,}250} \times \frac{4 \times 16.7}{\left[1 + 0.83\left(\dfrac{12}{25}\right)^2\right]} = 0.50$$

Therefore,

$$T_a = \frac{0.0019}{\sqrt{0.50}} \times 12 = 0.03 \text{ s}$$

In S.I.:

$$A_B = 22.9 \times 45.7 = 1{,}046.5 \text{ m}^2$$

$$A_i = (203/1{,}000) \times 7.6 = 1.55 \text{ m}^2$$

$$D_i = 7.6 \text{ m}$$

$$C_w = \frac{100}{1{,}046.5} \times \frac{4 \times 1.55}{\left[1 + 0.83\left(\dfrac{3.7}{7.6}\right)^2\right]} = 0.50$$

Therefore,

$$T_a = \frac{0.0058}{\sqrt{0.50}} \times 3.7 = 0.03 \text{ s}$$

The period of the combined structure, which is obtained from a dynamic analysis of the entire structure, is equal to 0.26 s, which is equal to 1.1 times the period of the upper structure (that is, $1.1 \times 0.24 = 0.26$ s).

c. The flexible upper portion must be designed as a separate structure using the appropriate value of R and ρ.

For the ordinary reinforced concrete moment frame (system C7): $R = 3$, $\Omega_0 = 3$, and $C_d = 2.5$ with no height limit for SDC B (ASCE/SEI Table 12.2-1).

For SDC B, $\rho = 1.0$ (ASCE/SEI 12.3.4.1).

d. The lower rigid portion must be designed as a separate structure using the appropriate value of R and ρ. Amplified reactions from the upper portion are applied to the lower portion where the amplification factor is equal to the ratio of R/ρ of the upper portion divided by R/ρ of the lower and must be greater than or equal to 1.0 (see Fig. 3.4).

For the ordinary reinforced concrete shear wall (system B5): $R = 5$, $\Omega_0 = 2.5$, and $C_d = 4.5$ with no height limit for SDC B (ASCE/SEI Table 12.2-1).

Amplification factor $= (3/1)/(5/1) = 0.6 < 1.0$, use 1.0

e. Both the top and bottom portions will be analyzed using the ELF procedure in ASCE/SEI 12.8.

Therefore, a two-stage equivalent lateral force procedure is permitted to be used.

Step 2—Determine the governing design coefficients

Design the upper ordinary reinforced concrete moment frames using $R = 3$, $\Omega_0 = 3$, and $C_d = 2.5$.

Design the lower ordinary reinforced concrete shear walls using $R = 5$, $\Omega_0 = 2.5$, and $C_d = 4.5$. This system is subjected to the combined effects of the amplified base shear, $V_{(upper)}$, from the upper portion and the lateral force due to the base shear from the lower portion, $V_{(lower)}$ (see Fig. 3.29).

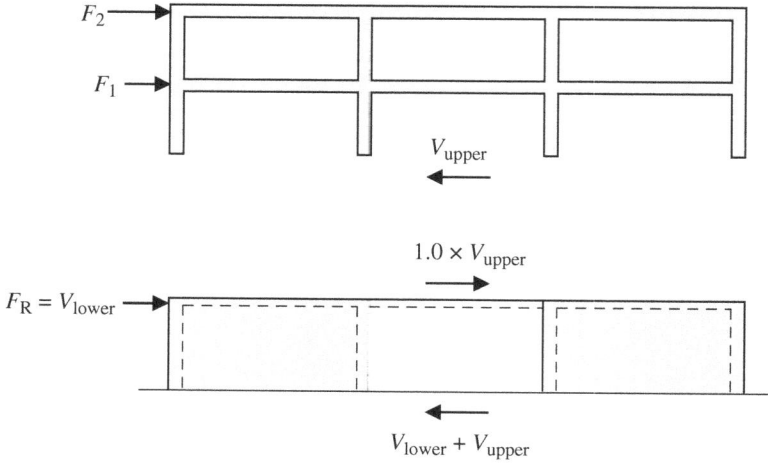

FIGURE 3.29   Combined seismic load effects on the lower portion of the building in Example 3.2.

3.16.3   Example 3.3—Combination of Framing Systems in the Same Direction: Horizontal Combinations

Determine the governing design coefficients R, Ω_0, and C_d in the north-south direction for the one-story warehouse building in Fig. 3.30 assuming the roof diaphragm is flexible. Assume the building is assigned to SDC C.

Solution

The requirements in ASCE/SEI 12.2.3.3 are used to determine the governing design coefficients.

   Step 1—Determine R, Ω_0, and C_d for each system   ASCE/SEI Table 12.2-1

   For the intermediate precast shear walls (system B8): $R = 5$, $\Omega_0 = 2.5$, and $C_d = 4.5$.
   For the steel ordinary concentrically braced frames (system B3): $R = 3.25$, $\Omega_0 = 2$, and $C_d = 3.25$.

   Step 2—Determine the governing design coefficients   ASCE/SEI 12.2.3.3

   Because of the horizontal combinations of SFRSs in the north-south direction, the requirements in ASCE/SEI 12.2.3.3 are applicable.
   Check if the conditions in the exception in ASCE/SEI 12.2.3.3 are satisfied:

- The building must be assigned to Risk Category I or II.
  Because of the warehouse occupancy, the building is assigned to Risk Category II in accordance with ASCE/SEI Table 1.5-1.

- The building must be no more than two stories above the grade plane.
  The building in this example is one story above the grade plane.

- The building must be of light-frame construction or must have flexible diaphragms.
  The roof diaphragm is given as flexible.

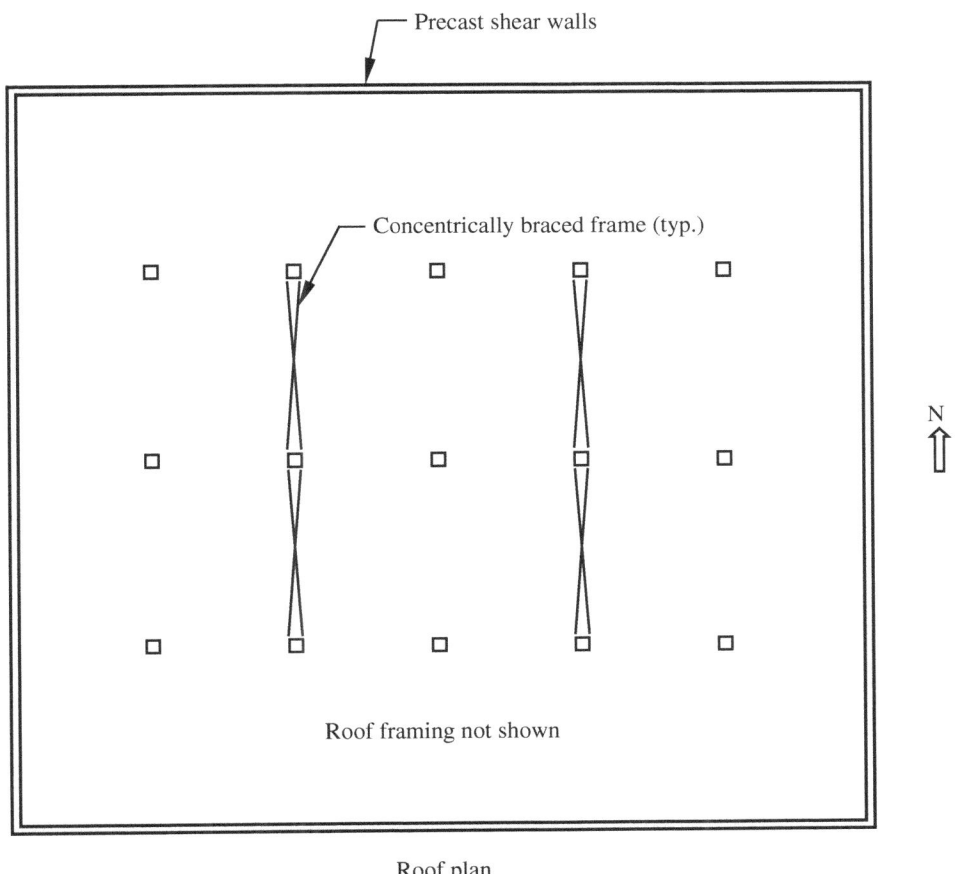

FIGURE 3.30 Horizontal combination of framing systems for the building in Example 3.3.

Therefore, all the conditions in the exception are satisfied and the intermediate precast shear walls are permitted to be designed using $R = 5$ and the steel ordinary concentrically braced frames are permitted to be designed using $R = 3.25$. The diaphragm must be designed using the least value of R in that direction, that is, using $R = 3.25$.

If the diaphragm was rigid instead of flexible, all conditions of the exception are not satisfied and the precast shear walls and the steel ordinary concentrically braced frames must both be designed using $R = 3.25$.

3.16.4 Example 3.4—Horizontal Irregularities

Determine whether the 10-story residential building in Fig. 3.31 possesses any of the horizontal structural irregularities in ASCE/SEI Table 12.3-1. Design data are given in Table 3.20.

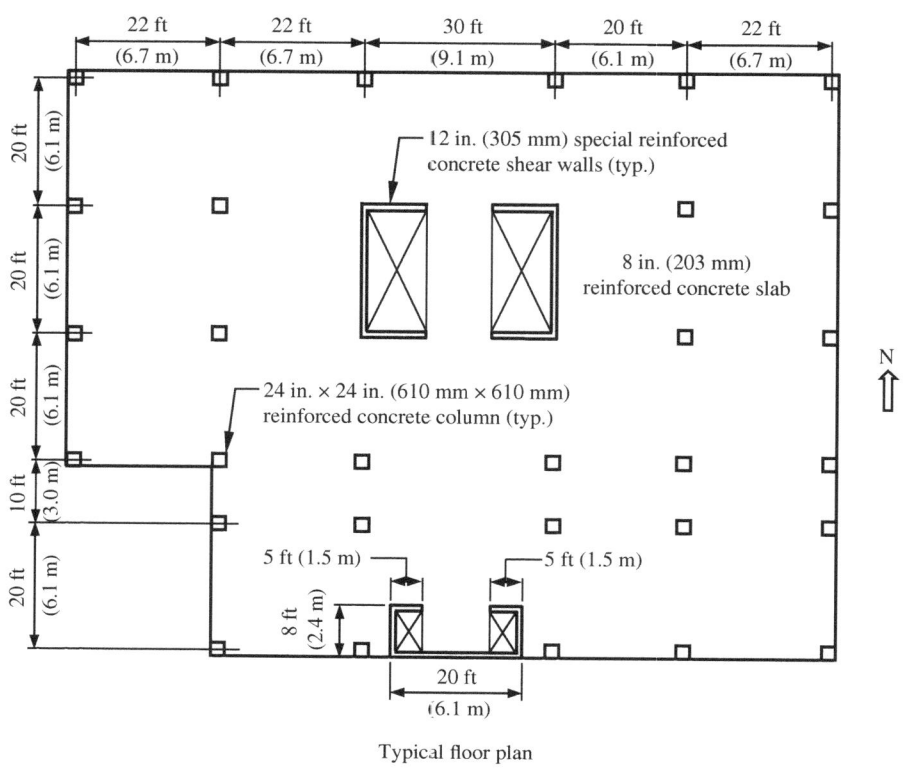

Typical floor plan

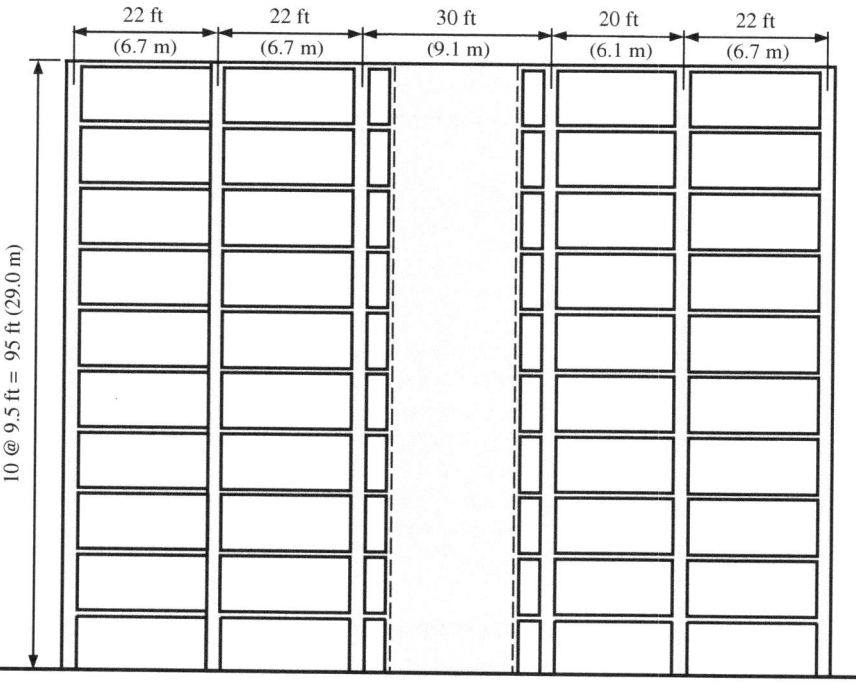

South elevation

Figure 3.31 Typical floor plan and elevation for the 10-story residential building in Example 3.4.

Location	Latitude: 38.602°, Longitude: −90.203°
Soil classification	Site class D (default)
Occupancy	Residential occupancy where less than 300 people congregate in one area
Material	Cast-in-place, reinforced concrete
SFRS	Building frame with special reinforced concrete shear walls (system B4)

TABLE 3.20 Design Data for the 10-Story Residential Building in Example 3.4

Solution

Step 1—Determine if a torsional irregularity or an extreme torsional irregularity exists

Story drifts are needed to ascertain whether torsional irregularity or an extreme torsional irregularity exists. The ELF procedure in ASCE/SEI 12.8 is sufficiently accurate to determine the displacements and story drifts for this purpose. The flowchart in Fig. 3.20 is used to determine the seismic base shear, V, and the lateral seismic forces, F_x, at the floor and roof levels.

Step 1a—Determine S_S and S_1   ASCE/SEI 11.4.2

In lieu of ASCE/SEI Figures 22-1 through 22-8, S_S and S_1 are determined from Refs. 5, 6, or 7:

$$S_S = 0.459$$
$$S_1 = 0.162$$

The structure is not permitted to be assigned to SDC A because $S_S > 0.15$ and $S_1 > 0.04$ (ASCE/SEI 11.4.2).

Step 1b—Determine the site class   ASCE/SEI 11.4.3

The site class is given as D (default) in Table 3.20.

Step 1c—Determine S_{MS} and S_{M1}   ASCE/SEI 11.4.4

Site coefficients F_a and F_v are obtained from ASCE/SEI Tables 11.4-1 and 11.4-2, respectively.
For site class D and $0.25 < S_S = 0.459 < 0.50$, $F_a = 1.43$ by linear interpolation.
For site class D and $0.10 < S_1 = 0.162 < 0.20$, $F_v = 2.28$ by linear interpolation.
Therefore,

$$S_{MS} = F_a S_S = 1.43 \times 0.459 = 0.656$$
$$S_{M1} = F_v S_1 = 2.28 \times 0.162 = 0.369$$

Step 1d—Determine S_{DS} and S_{D1}   ASCE/SEI 11.4.5

$$S_{DS} = \frac{2}{3} S_{MS} = \frac{2}{3} \times 0.656 = 0.437$$
$$S_{D1} = \frac{2}{3} S_{M1} = \frac{2}{3} \times 0.369 = 0.246$$

Step 1e—Determine the SDC   ASCE/SEI 11.6

The SDC is determined using Fig. 2.6 of this publication.

For a residential occupancy where less than 300 people congregate in one area, the Risk Category is II (see IBC Table 1604.5 and ASCE/SEI Table 1.5-1).

Because $S_1 = 0.162 < 0.75$, the building is not assigned to SDC E or F.

Check if all four conditions of ASCE/SEI 11.6 are satisfied:

- Check if $T_a < 0.8T_S$

Use ASCE/SEI Equation (12.8-7) with approximate period parameters for "other structural systems":

$$T_a = C_t h_n^x = 0.02 \times (95.0)^{0.75} = 0.61 \text{ s}$$
$$T_S = S_{D1}/S_{DS} = 0.246/0.437 = 0.56 \text{ s}$$
$$T_a = 0.61 \text{ s} > 0.8T_S = 0.45 \text{ s}$$

Because this condition is not satisfied, the SDC cannot be determined by ASCE/SEI Table 11.6-1 alone (ASCE/SEI 11.6).

Determine the SDC from ASCE/SEI Tables 11.6-1 and 11.6-2:

From ASCE/SEI Table 11.6-1: with $0.33 < S_{DS} = 0.437 < 0.50$, the SDC is C.

From ASCE/SEI Table 11.6-2: with $S_{D1} = 0.246 > 0.20$, the SDC is D.

Therefore, the SDC is D for this building.

Step 1f—Determine the response modification coefficient, R

For a building frame system with special reinforced concrete shear walls, $R = 6$ from ASCE/SEI Table 12.2-1.

Step 1g—Determine the importance factor, I_e

For a Risk Category II building, $I_e = 1.0$ from ASCE/SEI Table 1.5-2.

Step 1h—Determine the fundamental period of the building, T

A three-dimensional model of the building was constructed in accordance with ASCE/SEI 12.7 using Ref. 10 assuming rigid diaphragms at all levels. From the analysis, it was found that $T = 1.22$ s in the north-south direction and $T = 1.75$ s in the east-west direction.

The maximum permitted period, T_{max}, is determined in accordance with ASCE/SEI 12.8.2:

$$T_{max} = C_u T_a$$

The coefficient for upper limit on calculated period, C_u, is determined by ASCE/SEI Table 12.8-1. For $0.2 < S_{D1} = 0.246 < 0.3$, $C_u = 1.45$ by linear interpolation.

Therefore, $T_{max} = 1.45 \times 0.61 = 0.89$ s where T_a is determined in step 1e.

Because $T > T_{max}$, use $T = 0.89$ s in the determination of the seismic forces.

Step 1i—Determine the long-period transition period, T_L

From Refs. 5, 6, or 7, $T_L = 12$ s.

Step 1j—Determine the seismic response coefficient, C_s   ASCE/SEI 12.8.1.1

Determine C_s by ASCE/SEI Equation (12.8-2):

$$C_s = \frac{S_{DS}}{\left(\frac{R}{I_e}\right)} = \frac{0.437}{\left(\frac{6}{1.0}\right)} = 0.073$$

Because $T = 0.89$ s $< T_L = 12$ s, C_s need not exceed that determined by ASCE/SEI Equation (12.8-3):

$$C_s = \frac{S_{D1}}{T\left(\frac{R}{I_e}\right)} = \frac{0.246}{0.89 \times \left(\frac{6}{1.0}\right)} = 0.046$$

Minimum C_s from ASCE/SEI Equation (12.8-5) governs because $S_1 < 0.6$:

$$C_s = \text{larger of} \begin{cases} 0.044 S_{DS} I_e = 0.044 \times 0.437 \times 1.0 = 0.019 \\ 0.01 \end{cases}$$

Therefore, $C_s = 0.046$.

Step 1k—Determine the effective seismic weight, W　　　　　ASCE/SEI 12.7.2

The effective seismic weight, W, is the summation of the effective dead loads at each level and includes the weight of the structural members, superimposed dead loads, the weight of the cladding, and the weight of the mechanical equipment on the roof (details of the weight calculations are not shown here). The story weights are given in Table 3.21.

Step 1l—Determine the base shear, V　　　　　ASCE/SEI Equation (12.8-1)

$$V = C_s W = 0.046 \times 12,690 = 583.7 \text{ kips}$$

Level	Story Weight, w_x (kips)	Height, h_x (ft)	$w_x h_x^k$	Lateral Force, F_x (kips)	Story Shear, V_x (kips)
R	1,116	95.0	263,592	103.0	103.0
10	1,286	85.5	267,670	104.5	207.5
9	1,286	76.0	232,390	90.8	298.3
8	1,286	66.5	197,983	77.3	375.6
7	1,286	57.0	164,547	64.3	439.9
6	1,286	47.5	132,213	51.6	491.5
5	1,286	38.0	101,154	39.5	531.0
4	1,286	28.5	71,623	28.0	559.0
3	1,286	19.0	44,030	17.2	576.2
2	1,286	9.5	19,165	7.5	583.7
Σ	12,690		1,494,367	583.7	

1 ft = 0.3048 m; 1 kip = 4.4482 kN.

TABLE 3.21 Seismic Forces and Seismic Story Shears for the Residential Building in Example 3.4

In S.I.:

$$V = C_s W = 0.046 \times 56,448 = 2,596.6 \text{ kN}$$

Step 1m—Determine the exponent related to the structure period, k

ASCE/SEI 12.8.3

Because $0.5 \text{ s} < T = 0.89 \text{ s} < 2.5 \text{ s}$, $k = 0.75 + 0.5T = 1.20$.

Step 1n—Determine the lateral seismic force, F_x, at each level   ASCE/SEI 12.8.3

The force F_x is determined by ASCE/SEI Equations (12.8-11) and (12.8-12). A summary of the lateral seismic forces, F_x, and the story shears, V_x, are given in Table 3.21.

For example, at the roof level:

$$C_{vx} = \frac{w_x h_x^k}{\sum w_i h_i} = \frac{1,116 \times (95.0)^{1.2}}{1,494,367} = 0.1764$$

$$F_x = C_{vx} V = 0.1764 \times 583.7 = 103.0 \text{ kips}$$

In S.I.:

$$C_{vx} = \frac{4,964.2 \times (29.0)^{1.2}}{1,597,569} = 0.1767$$

$$F_x = 0.1767 \times 2,596.6 = 458.8 \text{ kN}$$

Step 1o—Determine the elastic displacements, δ_{xe}, at each level and check the conditions for torsional irregularities

The building was analyzed in the north-south direction using the lateral seismic forces in Table 3.21 assuming rigid diaphragms at all levels and including reduced stiffness properties due to cracking of the concrete (ASCE/SEI 12.7.3). The forces are applied at the CM at each level, which is displaced in each orthogonal direction 5 percent of the dimension of the structure perpendicular to the direction of the applied forces; this accidental torsion, M_{ta}, which is in addition to the inherent torsion, M_t, due to the CM and CR being at different locations, must be applied when determining if a horizontal irregularity exists (ASCE/SEI 12.8.4.2).

A type 1a torsional irregularity occurs where the maximum story drift, Δ_{max}, including accidental torsional effects, at one end of the structure transverse to an axis is more than 1.2 times the average of the story drifts, Δ_{avg}, of the two ends of the structure (see Table 3.6):

$$\Delta_{max} > 1.2\Delta_{avg} = \frac{1.2(\Delta_{max,x} + \Delta_{min,x})}{2}$$

The story drift at one end of the building is equal to the elastic displacement, δ_{xe}, due to the code-prescribed lateral forces at that level at that end minus the elastic displacement at the level below. A summary of the displacements and drifts at both ends of the structure at all levels is given in Table 3.22 for seismic forces in the north-south direction.

Story	$(\delta_{xe})_{max}$, (in.)	$\Delta_{max,x}$ (in.)	$(\delta_{xe})_{min}$, (in.)	$\Delta_{min,x}$ (in.)	Δ_{avg} (in.)	$\Delta_{max,x}/\Delta_{avg}$
10	1.72	0.20	0.93	0.12	0.16	1.25
9	1.52	0.21	0.81	0.12	0.17	1.24
8	1.31	0.21	0.69	0.12	0.17	1.24
7	1.10	0.21	0.57	0.12	0.17	1.24
6	0.89	0.21	0.45	0.11	0.16	1.31
5	0.68	0.20	0.34	0.10	0.15	1.33
4	0.48	0.17	0.24	0.09	0.13	1.31
3	0.31	0.15	0.15	0.07	0.11	1.36
2	0.16	0.11	0.08	0.05	0.08	1.38
1	0.05	0.05	0.03	0.03	0.04	1.25

1 in. = 25.4 mm.

TABLE 3.22 Lateral Displacements and Story Drifts in the North-South Direction for the Residential Building in Example 3.4

For example, in the 10th story:

$$\Delta_{max,x} = 1.72 - 1.52 = 0.20 \text{ in.}$$
$$\Delta_{min,x} = 0.93 - 0.81 = 0.12 \text{ in.}$$

$$\Delta_{avg} = \frac{0.20 + 0.12}{2} = 0.16 \text{ in.}$$
$$\Delta_{max,x}/\Delta_{avg} = 0.20/0.16 = 1.25$$

In S.I.:

$$\Delta_{max,x} = 43.7 - 38.6 = 5.1 \text{ mm}$$
$$\Delta_{min,x} = 23.6 - 20.6 = 3.0 \text{ mm}$$

$$\Delta_{avg} = \frac{5.1 + 3.0}{2} = 4.1 \text{ mm}$$
$$\Delta_{max,x}/\Delta_{avg} = 5.1/4.1 = 1.24$$

It is evident from Table 3.22 that a type 1a torsional irregularity exists in all stories in the north-south direction because the ratio of $\Delta_{max,x}$ to Δ_{avg} is greater than 1.2. A type 1b extreme torsional irregularity does not exist because the ratio is less than 1.4 in all stories.

According to ASCE/SEI 12.8.4.3, the accidental torsional moments, M_{ta}, must be increased by the torsional amplification factor, A_x, determined by ASCE/SEI Equation (12.8-14) where a torsional irregularity exists:

$$A_x = \left[\frac{(\delta_{xe})_{max}}{1.2(\delta_{xe})_{avg}}\right]^2$$

For example, in the 10th story:

$$A_x = \left[\frac{1.72}{1.2 \times \left(\frac{1.72 + 0.93}{2} \right)} \right]^2 = 1.17 > 1.0$$

In S.I.:

$$A_x = \left[\frac{43.7}{1.2 \times \left(\frac{43.7 + 23.6}{2} \right)} \right]^2 = 1.17 > 1.0$$

Step 2—Determine if a reentrant corner irregularity exists

According to ASCE/SEI Table 12.3-1, a reentrant corner irregularity exists where both plan dimensions of the structure beyond a reentrant corner are greater than 15 percent of the plan dimension of the structure in the given direction (see Table 3.6).

$$L_{xp} = 22.0 \text{ ft} > 0.15 L_x = 0.15 \times 118.0 = 17.7 \text{ ft}$$
$$L_{yp} = 30.0 \text{ ft} > 0.15 L_y = 0.15 \times 92.0 = 13.8 \text{ ft}$$

Therefore, a reentrant corner irregularity exists at all levels.

Step 3—Determine if a diaphragm discontinuity irregularity exists

A diaphragm discontinuity does not exist because the diaphragms at each level do not have an abrupt discontinuity or variation in stiffness. Also, the total area of openings in the diaphragms, which are consistent with opening sizes for elevators and stairs, is less than 50 percent of the gross enclosed diaphragm area.

Step 4—Determine if an out-of-plane offset irregularity exists

An out-of-plane offset irregularity does not exist because there are no out-of-plane offsets of the shear walls.

Step 5—Determine if a nonparallel system irregularity exists

A nonparallel system irregularity does not exist because all the shear walls are parallel to a major orthogonal axis of the building.

3.16.5 Example 3.5—Vertical Irregularities

Determine whether the four-story emergency treatment facility in Fig. 3.32 possesses any of the vertical structural irregularities in ASCE/SEI Table 12.3-2. Design data are given in Table 3.23.

Solution

Step 1—Determine if a stiffness-soft story irregularity exists

Lateral story stiffnesses are needed to determine whether a soft story irregularity exists. Story stiffness can be obtained by dividing the story shear by the story drift in the direction of analysis. Therefore, the lateral displacements at

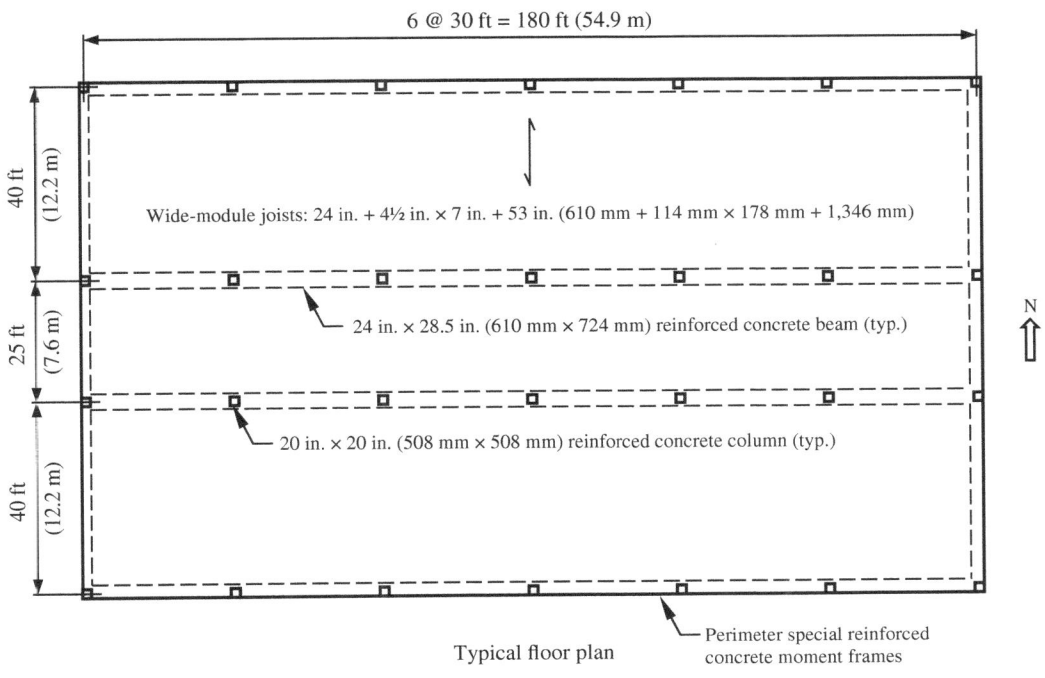

Typical floor plan

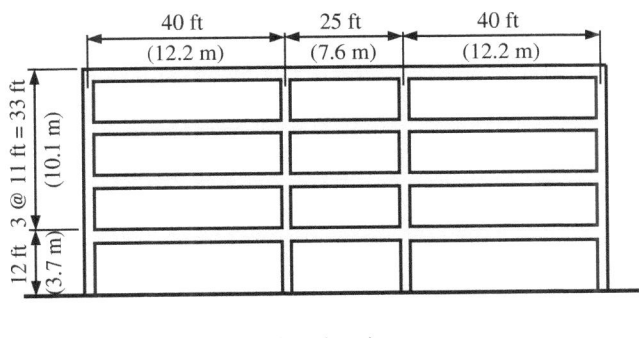

East elevation

FIGURE 3.32 Typical floor plan and elevation for the four-story emergency treatment facility in Example 3.5.

the CM for each story are needed. The ELF procedure in ASCE/SEI 12.8 is sufficiently accurate to determine the displacements and story drifts for this purpose. The flowchart in Fig. 3.20 is used to determine the seismic base shear, V, and the lateral seismic forces, F_x, at the floor and roof levels, which will be used to determine the lateral displacements.

Location	Latitude: 37.145°, Longitude: –93.278°
Soil classification	Site class D (stiff soil)
Occupancy	Essential facility
Material	Cast-in-place, reinforced concrete
SFRS	Special reinforced concrete moment frames (system C5)

TABLE 3.23 Design Data for the Four-Story Emergency Treatment Facility in Example 3.5

Step 1a—Determine S_S and S_1   ASCE/SEI 11.4.2

In lieu of ASCE/SEI Figures 22-1 through 22-8, S_S and S_1 are determined from Refs. 5, 6, or 7:

$$S_S = 0.192$$
$$S_1 = 0.106$$

The structure is not permitted to be assigned to SDC A because $S_S > 0.15$ and $S_1 > 0.04$ (ASCE/SEI 11.4.2).

Step 1b—Determine the site class   ASCE/SEI 11.4.3

The site class is given as D (stiff soil) in Table 3.23.

Step 1c—Determine S_{MS} and S_{M1}   ASCE/SEI 11.4.4

Site coefficients F_a and F_v are obtained from ASCE/SEI Tables 11.4-1 and 11.4-2, respectively.

For site class D and $S_S = 0.192 < 0.25$, $F_a = 1.60$.

For site class D and $0.10 < S_1 = 0.106 < 0.20$, $F_v = 2.39$ by linear interpolation.

Therefore,

$$S_{MS} = F_a S_S = 1.60 \times 0.192 = 0.307$$

$$S_{M1} = F_v S_1 = 2.39 \times 0.106 = 0.253$$

Step 1d—Determine S_{DS} and S_{D1}   ASCE/SEI 11.4.5

$$S_{DS} = \frac{2}{3} S_{MS} = \frac{2}{3} \times 0.307 = 0.205$$

$$S_{D1} = \frac{2}{3} S_{M1} = \frac{2}{3} \times 0.253 = 0.169$$

Step 1e—Determine the SDC   ASCE/SEI 11.6

The SDC is determined using Fig. 2.6 of this publication.

For an essential facility, the Risk Category is IV (see IBC Table 1604.5 and ASCE/SEI Table 1.5-1).

Because $S_1 = 0.106 < 0.75$, the building is not assigned to SDC E or F.

Determine the SDC from ASCE/SEI Tables 11.6-1 and 11.6-2 (the four conditions in ASCE/SEI 11.6 to determine if the SDC can be obtained from ASCE/SEI Table 11.6-1 alone are not checked in this example):

From ASCE/SEI Table 11.6-1: with $0.167 < S_{DS} = 0.205 < 0.33$, the SDC is C.
From ASCE/SEI Table 11.6-2: with $0.133 < S_{D1} = 0.169 < 0.20$, the SDC is D.
Therefore, the SDC is D for this building.

Step 1f—Determine the response modification coefficient, R

For special reinforced concrete moment frames, $R = 8$ from ASCE/SEI Table 12.2-1.

Step 1g—Determine the importance factor, I_e

For a Risk Category IV building, $I_e = 1.5$ from ASCE/SEI Table 1.5-2.

Step 1h—Determine the fundamental period of the building, T

A three-dimensional model of the building was constructed in accordance with ASCE/SEI 12.7 using Ref. 10 assuming rigid diaphragms at all levels. From the analysis, it was found that $T = 1.70$ s in the north-south direction and $T = 1.40$ s in the east-west direction.

The maximum permitted period, T_{max}, is determined in accordance with ASCE/SEI 12.8.2:

$$T_{max} = C_u T_a$$

The coefficient for upper limit on calculated period, C_u, is determined by ASCE/SEI Table 12.8-1. For $0.15 < S_{D1} = 0.169 < 0.20$, $C_u = 1.56$ by linear interpolation.

Use ASCE/SEI Equation (12.8-7) with approximate period parameters for "concrete moment-resisting frames":

$$T_a = C_t h_n^x = 0.016 \times (45.0)^{0.9} = 0.49 \text{ s}$$

In S.I.:

$$T_a = 0.0466 \times (13.72)^{0.9} = 0.49 \text{ s}$$

Therefore, $T_{max} = 1.56 \times 0.49 = 0.76$ s.

Because $T > T_{max}$, use $T = 0.76$ s in the determination of the seismic forces in both orthogonal directions.

Step 1i—Determine the long-period transition period, T_L

From Refs. 5, 6, or 7, $T_L = 12$ s.

Step 1j—Determine the seismic response coefficient, C_s   ASCE/SEI 12.8.1.1

Determine C_s by ASCE/SEI Equation (12.8-2):

$$C_s = \frac{S_{DS}}{\left(\dfrac{R}{I_e}\right)} = \frac{0.205}{\left(\dfrac{8}{1.5}\right)} = 0.038$$

Because $T = 0.76$ s $< T_L = 12$ s, C_s need not exceed that determined by ASCE/SEI Equation (12.8-3):

$$C_s = \frac{S_{D1}}{T\left(\dfrac{R}{I_e}\right)} = \frac{0.169}{0.76 \times \left(\dfrac{8}{1.5}\right)} = 0.042$$

Minimum C_s from ASCE/SEI Equation (12.8-5) governs because $S_1 < 0.6$:

$$C_s = \text{larger of } \begin{cases} 0.044S_{DS}I_e = 0.044 \times 0.205 \times 1.5 = 0.014 \\ 0.01 \end{cases}$$

Therefore, $C_s = 0.038$.

Step 1k—Determine the effective seismic weight, W ASCE/SEI 12.7.2

The effective seismic weight, W, is the summation of the effective dead loads at each level and includes the weight of the structural members, superimposed dead loads, the weight of the cladding, and the weight of the mechanical equipment on the roof (details of the weight calculations are not shown here).

The story weights are given in Table 3.24.

Step 1l—Determine the base shear, V ASCE/SEI Equation (12.8-1)

$$V = C_sW = 0.038 \times 10,447 = 397.0 \text{ kips}$$

In S.I.:

$$V = C_sW = 0.038 \times 46,470 = 1,765.9 \text{ kN}$$

Step 1m—Determine the exponent related to the structure period, k

ASCE/SEI 12.8.3

Because $0.5 \ s < T = 0.76 \ s < 2.5 \ s$, $k = 0.75 + 0.5T = 1.13$.

Step 1n—Determine the lateral seismic force, F_x, at each level ASCE/SEI 12.8.3

The force F_x is determined by ASCE/SEI Equations (12.8-11) and (12.8-12). A summary of the lateral seismic forces, F_x, and the story shears, V_x, are given in Table 3.24.

For example, at the roof level:

$$C_{vx} = \frac{w_x h_x^k}{\sum w_i h_i} = \frac{2,561 \times (45.0)^{1.13}}{464,531} = 0.4069$$

$$F_x = C_{vx}V = 0.4069 \times 397.0 = 161.6 \text{ kips}$$

Level	Story Weight, w_x (kips)	Height, h_x (ft)	$w_x h_x^k$	Lateral Force, F_x (kips)	Story Shear, V_x (kips)
R	2,561	45.0	189,034	161.6	161.6
4	2,623	34.0	141,049	120.5	282.1
3	2,623	23.0	90,688	77.5	359.6
2	2,640	12.0	43,760	37.4	397.0
Σ	10,447		464,531	397.0	

1 ft = 0.3048 m; 1 kip = 4.4482 kN.

TABLE 3.24 Seismic Forces and Seismic Story Shears for the Essential Facility in Example 3.5

In S.I.:

$$C_{vx} = \frac{11{,}392.6 \times (13.7)^{1.13}}{539{,}724} = 0.4069$$

$$F_x = 0.4069 \times 1{,}765.9 = 718.6 \text{ kN}$$

Step 1o—Determine the elastic displacements, δ_{xe}, at the CM at each level and check if a stiffness-soft story irregularity exists

The building was analyzed in the north-south and east-west directions using the lateral seismic forces in Table 3.24 assuming rigid diaphragms at all levels and including reduced stiffness properties due to cracking of the concrete (ASCE/SEI 12.7.3). The forces are applied at the CM at each level. Because the analysis is not checking for horizontal irregularities, the CM need not be displaced in each orthogonal direction 5 percent of the dimension of the structure perpendicular to the direction of the applied forces (ASCE/ SEI 12.8.4.2). A summary of the elastic displacements, δ_{xe}, and the story drifts, Δ_x, due to the code-prescribed forces is given in Table 3.25.

Story stiffness, k_x, is determined by dividing the story shear, V_x, from Table 3.24 by the story drift, Δ_x, from Table 3.25. A summary of the story stiffnesses in the north-south and east-west directions is given in Table 3.26.

	North-South		East-West	
Story	δ_{xe}, **in. (mm)**	Δ_x, **in. (mm)**	δ_{xe}, **in. (mm)**	Δ_x, **in. (mm)**
4	1.52 (38.6)	0.21 (5.3)	1.04 (26.4)	0.15 (3.8)
3	1.31 (33.3)	0.35 (8.9)	0.89 (22.6)	0.25 (6.3)
2	0.96 (24.4)	0.45 (11.4)	0.64 (16.3)	0.31 (7.9)
1	0.51 (13.0)	0.51 (13.0)	0.33 (8.4)	0.33 (8.4)

TABLE 3.25 Lateral Displacements and Story Drifts in the North-South and East-West Directions for the Essential Facility in Example 3.5

		North-South		East-West	
Story	V_x, **kips (kN)**	Δ_x, **in. (mm)**	k_x, **kips/in. (kN/mm)**	Δ_x, **in. (mm)**	k_x, **kips/in. (kN/mm)**
4	161.6 (718.8)	0.21 (5.3)	769.5 (135.6)	0.15 (3.8)	1,077.3 (189.2)
3	282.1 (1,254.8)	0.35 (8.9)	806.0 (141.0)	0.25 (6.3)	1,128.4 (199.2)
2	359.6 (1,599.6)	0.45 (11.4)	799.1 (140.3)	0.31 (7.9)	1,160.0 (202.5)
1	397.0 (1,765.9)	0.51 (13.0)	778.4 (135.8)	0.33 (8.4)	1,203.0 (210.2)

TABLE 3.26 Story Stiffnesses in the North-South and East-West Directions for the Essential Facility in Example 3.5

A type 1a soft story exits in the first story where one of the two following conditions is satisfied:

$$k_1 < 0.7k_2$$

$$k_1 < 0.8\left(\frac{k_2 + k_3 + k_4}{3}\right)$$

Check if a type 1a soft story irregularity exists in the first story:

- North-south direction

$$k_1 = 778.4 \text{ kips/in.} > 0.7k_2 = 0.7 \times 799.1 = 559.4 \text{ kips/in.}$$

$$k_1 = 778.4 \text{ kips/in.}$$

$$> 0.8\left(\frac{k_2 + k_3 + k_4}{3}\right) = 0.8 \times \left(\frac{799.1 + 806.0 + 769.5}{3}\right) = 633.2 \text{ kips/in.}$$

In S.I.:

$$k_1 = 135.8 \text{ kN/mm} > 0.7k_2 = 0.7 \times 140.3 = 98.2 \text{ kN/mm}$$

$$k_1 = 135.8 \text{ kN/mm}$$

$$> 0.8\left(\frac{k_2 + k_3 + k_4}{3}\right) = 0.8 \times \left(\frac{140.3 + 141.0 + 135.6}{3}\right) = 111.2 \text{ kN/mm}$$

Therefore, a type 1a soft story irregularity does not exist in the first story for analysis in the north-south direction.

- East-west direction

$$k_1 = 1,203.0 \text{ kips/in.} > 0.7k_2 = 0.7 \times 1,160.0 = 812.0 \text{ kips/in.}$$

$$k_1 = 1,203.0 \text{ kips/in.}$$

$$> 0.8\left(\frac{k_2 + k_3 + k_4}{3}\right) = 0.8 \times \left(\frac{1,160.0 + 1,128.4 + 1,077.3}{3}\right) = 897.5 \text{ kips/in.}$$

In S.I.:

$$k_1 = 210.2 \text{ kN/mm} > 0.7k_2 = 0.7 \times 202.5 = 141.8 \text{ kN/mm}$$

$$k_1 = 210.2 \text{ kN/mm}$$

$$> 0.8\left(\frac{k_2 + k_3 + k_4}{3}\right) = 0.8 \times \left(\frac{202.5 + 199.2 + 189.2}{3}\right) = 157.6 \text{ kN/mm}$$

Therefore, a type 1a soft story irregularity does not exist in the first story for analysis in the east-west direction.

It is evident from the above calculations that a type 1b extreme soft story does not exist in the first story in either direction.

Step 2—Determine if a weight (mass) irregularity exists

It is evident from Table 3.24 that the effective mass of any story is not more than 150 percent of the effective mass of an adjacent story, so a weight (mass) irregularity does not exist.

Step 3—Determine if a vertical geometric irregularity exists

Because the horizontal dimensions of the SFRSs are the same over the height of the building, a vertical geometric irregularity does not exist.

Step 4—Determine if an in-plane discontinuity in vertical lateral force–resisting element irregularity exists

There are no in-plane offsets of the vertical SFRSs in the building, so this irregularity does not exist.

Step 5—Determine if discontinuity in lateral strength–weak story or extreme weak story irregularity exists

A weak story irregularity exists where a story lateral strength is less than 80 percent of that in the story above. The story lateral strength is the total lateral strength of all seismic-resisting elements sharing the story shear in the direction of analysis. Similarly, an extreme weak story exists where a story lateral strength is less than 65 percent of that in the story above.

Because the SFRSs consist of special moment frames, the sums of the column shear forces in the first and second stories are needed to determine if a weak story irregularity exists in the first story. The shear forces are calculated based on the nominal flexural strengths of the beams and columns in the special moment frames. Assuming the members have been designed such that a strong column-weak beam condition exists at the joints, the beams yield first, which means the column shear forces can be calculated using the nominal flexural strengths of the beams. The nominal negative and positive flexural strengths are equal to the following:

$$\left(M_n^-\right)_{beam} = 373.5 \text{ ft-kips (506.4 kN-m)}$$

$$\left(M_n^+\right)_{beam} = 297.8 \text{ ft-kips (403.8 kN-m)}$$

Check for a weak story irregularity in the north-south direction:

Free-body diagrams of edge and interior columns in the first story are given in Fig. 3.33. Because the columns are uniform in size over the height of the building, the beam nominal flexural strengths are distributed above and below the floor level in proportion to the inverse of the clear column heights. At the base of the columns, the bending moments are equal to the strength of the foundations supporting the columns, which are based on the bending moments transferred to the foundation from the lateral analysis. Similar free-body diagrams for the columns in the second floor are given in Fig. 3.34. Because the story heights for the second and third floor are the same, one-half of the beam flexural strengths are transferred to the columns.

The first- and second-story strengths are equal to the sum of the column shear forces in those stories:

$$\text{First-story strength} = 4 \times (33.2 + 55.7) = 355.6 \text{ kips}$$

$$\text{Second-story strength} = 4 \times (44.3 + 79.5) = 495.2 \text{ kips}$$

$$\frac{\text{First-story strength}}{\text{Second-story strength}} = \frac{355.6}{495.2} = 0.72 < 0.80$$

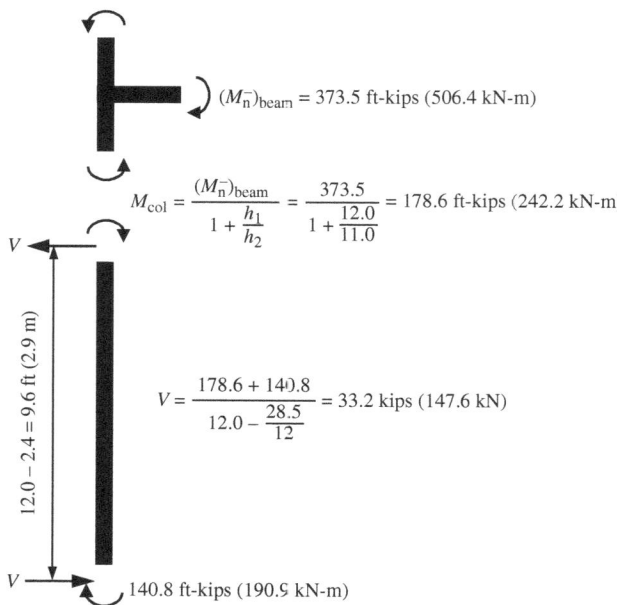

Edge column

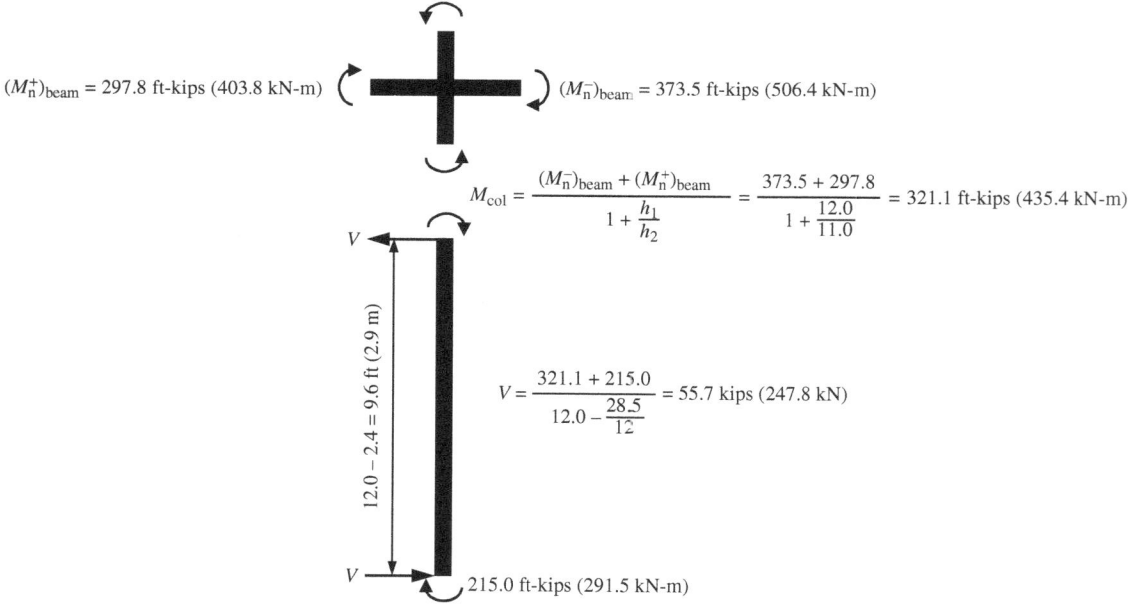

Interior column

FIGURE 3.33   Free-body diagrams of the first-story columns in the essential facility in Example 3.5.

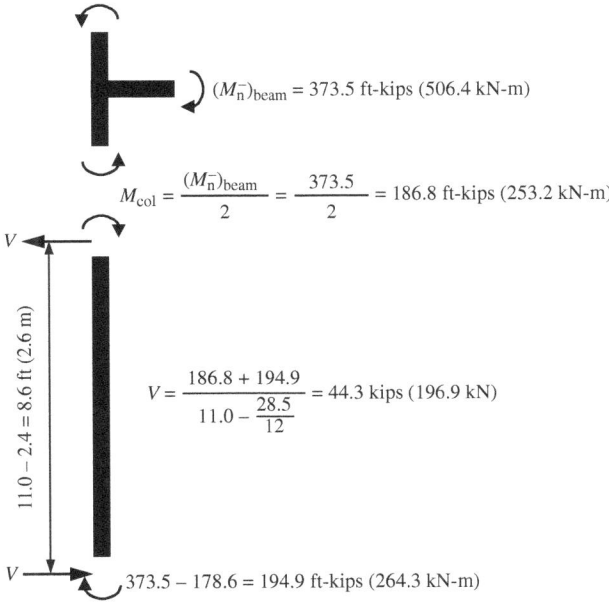

$(M_n^-)_{beam} = 373.5$ ft-kips (506.4 kN-m)

$M_{col} = \dfrac{(M_n^-)_{beam}}{2} = \dfrac{373.5}{2} = 186.8$ ft-kips (253.2 kN-m)

$11.0 - 2.4 = 8.6$ ft (2.6 m)

$V = \dfrac{186.8 + 194.9}{11.0 - \dfrac{28.5}{12}} = 44.3$ kips (196.9 kN)

$373.5 - 178.6 = 194.9$ ft-kips (264.3 kN-m)

Edge column

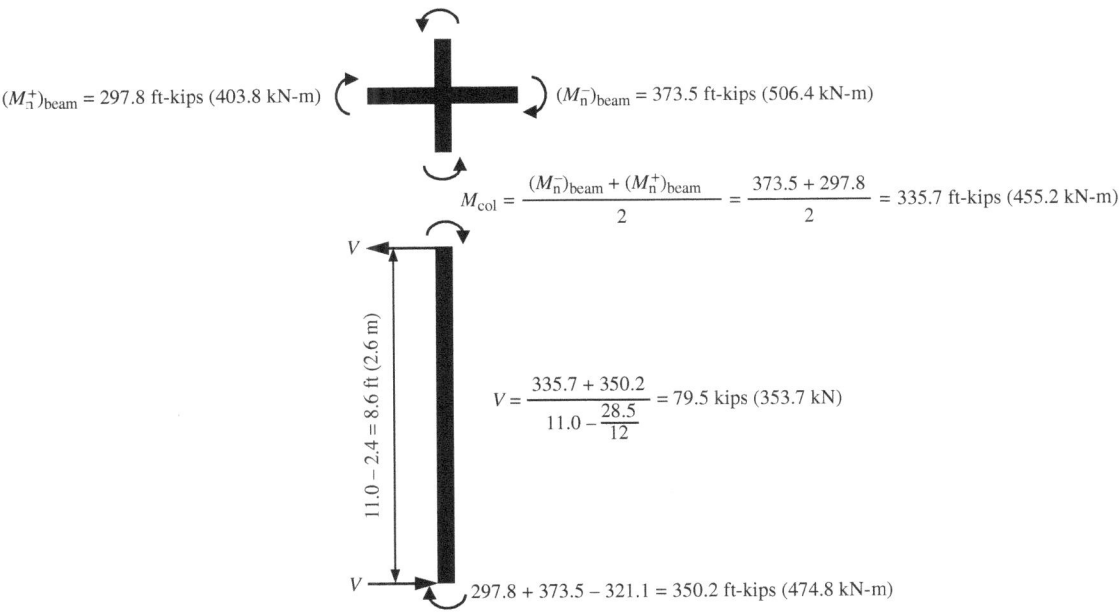

$(M_\cap^+)_{beam} = 297.8$ ft-kips (403.8 kN-m)

$(M_n^-)_{beam} = 373.5$ ft-kips (506.4 kN-m)

$M_{col} = \dfrac{(M_n^-)_{beam} + (M_n^+)_{beam}}{2} = \dfrac{373.5 + 297.8}{2} = 335.7$ ft-kips (455.2 kN-m)

$11.0 - 2.4 = 8.6$ ft (2.6 m)

$V = \dfrac{335.7 + 350.2}{11.0 - \dfrac{28.5}{12}} = 79.5$ kips (353.7 kN)

$297.8 + 373.5 - 321.1 = 350.2$ ft-kips (474.8 kN-m)

Interior column

FIGURE 3.34 Free-body diagrams of the second-story columns in the essential facility in Example 3.5.

In S.I.:

$$\text{First-story strength} = 4 \times (147.6 + 247.8) = 1,581.6 \text{ kN}$$

$$\text{Second-story strength} = 4 \times (194.9 + 353.7) = 2,194.4 \text{ kN}$$

$$\frac{\text{First-story strength}}{\text{Second-story strength}} = \frac{1,581.6}{2,194.4} = 0.72 < 0.80$$

Therefore, a weak story irregularity exists in the north-south direction. An extreme weak story does not exist because the ratio of the first story to second story strengths is greater than 0.65 (see ASCE/SEI Table 12.3-2).

In the east-west direction, the ratio of the first-story to second-story strengths is greater than 0.80, which means a weak story irregularity does not exist in that direction.

Structures assigned to SDC D with vertical irregularity type 5a (weak story irregularity) are permitted; structures assigned to SDC D with vertical irregularity type 5b (extreme weak story irregularity) are not permitted (ASCE/SEI 12.3.3.1).

3.16.6 Example 3.6—Redundancy Factor

Determine the redundancy factor, ρ, in the north-south and east-west directions for the 10-story residential building in Fig. 3.35. Assume the building is assigned to SDC D.

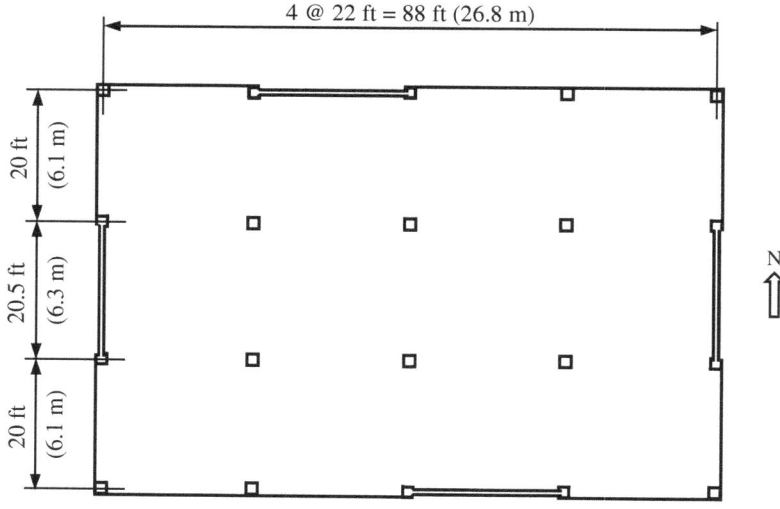

Story heights = 11 ft (3.4 m)

FIGURE 3.35 Typical floor plan for the residential building in Example 3.6.

Solution

Step 1—Determine ρ using ASCE/SEI 12.3.4.2(b)

It is permitted to take $\rho = 1.0$ for regular structures in plan with at least two bays of seismic force–resisting perimeter framing on each side of the structure in each orthogonal direction at each story resisting more than 35 percent of the base shear.

For structures with shear walls of other than light-frame construction, the number of bays is equal to the length of the shear wall divided by the story height:

In the north-south direction: Number of bays = 20.5/11.0 = 1.9 < 2.0

In the east-west direction: Number of bays = 22.0/11.0 = 2

In S.I.:

In the north-south direction: Number of bays = 6.25/3.35 = 1.9 < 2.0

In the east-west direction: Number of bays = 6.71/3.35 = 2

Therefore, it is permitted to take $\rho = 1.0$ in the east-west direction; in the north-south direction, $\rho = 1.3$ because the number of bays is less than 2.

Step 2—Determine ρ using ASCE/SEI 12.3.4.2(a)

For buildings with shear walls, it is permitted to take $\rho = 1.0$ if removal of a shear wall with a height-to-length ratio greater than 1.0 within a story would not result in more than a 33 percent reduction in story strength nor an extreme torsional irregularity (ASCE/SEI Table 12.3-3). Because the height-to-length ratios of the shear walls are less than 1.0, $\rho = 1.0$ in both directions based on the requirements in ASCE/SEI Table 12.3-3.

3.16.7   Example 3.7—ELF Procedure

Determine the seismic forces in the north-south direction for the 10-story residential building in Fig. 3.35 using the ELF procedure. Design data are given in Table 3.27.

Solution

Step 1—Determine S_S and S_1 ASCE/SEI 11.4.2

In lieu of ASCE/SEI Figures 22-1 through 22-8, S_S and S_1 are determined from Refs. 5, 6, or 7:

$$S_S = 1.952$$

Location	Latitude: 34.045°, Longitude: −118.258°
Soil classification	Site class D (default)
Occupancy	Residential occupancy where less than 300 people congregate in one area
Material	Cast-in-place, reinforced concrete
SFRS	Building frame with special reinforced concrete shear walls (system B4)

TABLE 3.27   Design Data for the 10-Story Residential Building in Example 3.7

$$S_1 = 0.694$$

The structure is not permitted to be assigned to SDC A because $S_S > 0.15$ and $S_1 > 0.04$ (ASCE/SEI 11.4.2).

Step 2—Determine the site class   ASCE/SEI 11.4.3

The site class is given as D (default) in Table 3.27.

Step 3—Determine S_{MS} and S_{M1}   ASCE/SEI 11.4.4

Site coefficients F_a and F_v are obtained from ASCE/SEI Tables 11.4-1 and 11.4-2, respectively.

For site class D and $S_S = 1.952 > 1.50$, $F_a = 1.20$.

For site class D and $S_1 = 0.694 > 0.60$, $F_v = 1.70$.

Because $S_1 > 0.20$ on a site class D site, a ground motion hazard analysis must be performed (ASCE/SEI 11.4.8(3)). However, it is assumed that exception 2 in ASCE/SEI 11.4.8 will be satisfied so that such an analysis is not required.

Therefore,

$$S_{MS} = F_a S_S = 1.20 \times 1.952 = 2.342$$
$$S_{M1} = F_v S_1 = 1.70 \times 0.694 = 1.180$$

Step 4—Determine S_{DS} and S_{D1}   ASCE/SEI 11.4.5

$$S_{DS} = \frac{2}{3} S_{MS} = \frac{2}{3} \times 2.342 = 1.561$$
$$S_{D1} = \frac{2}{3} S_{M1} = \frac{2}{3} \times 1.180 = 0.787$$

Step 5—Determine the SDC   ASCE/SEI 11.6

For a residential occupancy where less than 300 people congregate in one area, the risk category is II (see IBC Table 1604.5 and ASCE/SEI Table 1.5-1).

Because $S_1 = 0.694 < 0.75$, the building is not assigned to SDC E or F.

Determine the SDC from ASCE/SEI Tables 11.6-1 and 11.6-2:

From ASCE/SEI Table 11.6-1: with $S_{DS} = 1.561 > 0.50$, the SDC is D.

From ASCE/SEI Table 11.6-2: with $S_{D1} = 0.787 > 0.20$, the SDC is D.

Therefore, the SDC is D for this building.

Step 6—Determine the response modification coefficient, R

For a building frame system with special reinforced concrete shear walls, $R = 6$ from ASCE/SEI Table 12.2-1.

Step 7—Determine the importance factor, I_e

For a Risk Category II building, $I_e = 1.0$ from ASCE/SEI Table 1.5-2.

Step 8—Determine the fundamental period of the building, T

A three-dimensional model of the building was constructed in accordance with ASCE/SEI 12.7 using Ref. 10 assuming rigid diaphragms at all levels. From the analysis, it was found that $T = 0.77$ s in the north-south direction.

The maximum permitted period, T_{max}, is determined in accordance with ASCE/SEI 12.8.2:

$$T_{max} = C_u T_a$$

The coefficient for upper limit on calculated period, C_u, is determined by ASCE/SEI Table 12.8-1. For $S_{D1} = 0.787 > 0.40$, $C_u = 1.4$.

The approximate period, T_a, is determined by ASCE/SEI Equation (12.8-7) using the approximate period parameters from ASCE/SEI Table 12.8-2 for "other structural systems":

$$T_a = C_t h_n^x = 0.02 \times (110.0)^{0.75} = 0.68 \text{ s}$$

In S.I.:

$$T_a = 0.0488 \times (33.53)^{0.75} = 0.68 \text{ s}$$

Therefore, $T_{max} = 1.4 \times 0.68 = 0.95$ s.

Because $T < T_{max}$, use $T = 0.77$ s in the determination of the seismic forces in the north-south direction.

Step 9—Determine the long-period transition period, T_L

From Refs. 5, 6, or 7, $T_L = 8$ s.

Step 10—Determine the seismic response coefficient, C_s   ASCE/SEI 12.8.1.1

Determine C_s by ASCE/SEI Equation (12.8-2):

$$C_s = \frac{S_{DS}}{\left(\dfrac{R}{I_e}\right)} = \frac{1.561}{\left(\dfrac{6}{1.0}\right)} = 0.260$$

Because $T = 0.77$ s $< T_L = 8$ s, C_s need not exceed that determined by ASCE/SEI Equation (12.8-3):

$$C_s = \frac{S_{D1}}{T\left(\dfrac{R}{I_e}\right)} = \frac{0.787}{0.77 \times \left(\dfrac{6}{1.0}\right)} = 0.170$$

Because $S_1 > 0.6$, minimum C_s is equal to the greater of the values from ASCE/SEI Equations (12.8-5) and (12.8-6):

$$C_s = \text{larger of} \begin{cases} 0.044 S_{DS} I_e = 0.044 \times 1.561 \times 1.0 = 0.069 \\ 0.01 \\ 0.5 S_1 / (R/I_e) = 0.5 \times 0.694/(6/1.0) = 0.058 \end{cases}$$

Therefore, $C_s = 0.170$. However, it was assumed in step 3 that a ground motion hazard analysis in accordance with ASCE/SEI 21.2 was not required because the second exception in ASCE/SEI 11.4.8 is applicable. Therefore, determine C_s based on that exception:

$$1.5 T_S = 1.5 S_{D1}/S_{DS} = 1.5 \times 0.787/1.561 = 0.756 \text{ s} < T = 0.77 \text{ s} < T_L = 8 \text{ s}$$

Thus, C_s must be taken as 1.5 times the value computed in accordance with ASCE/SEI Equation (12.8-3):

$$C_s = 1.5 \times 0.170 = 0.255$$

Step 11—Determine the effective seismic weight, W   ASCE/SEI 12.7.2

The effective seismic weight, W, is the summation of the effective dead loads at each level and includes the weight of the structural members, superimposed dead loads, the weight of the cladding, and the weight of the mechanical equipment on the roof (details of the weight calculations are not shown here). The story weights are given in Table 3.28.

Step 12—Determine the base shear, V   ASCE/SEI Equation (12.8-1)

$$V = C_s W = 0.255 \times 8,016 = 2,044.1 \text{ kips}$$

In S.I.:

$$V = C_s W = 0.255 \times 35,657 = 9,092.5 \text{ kN}$$

Step 13—Determine the exponent related to the structure period, k

ASCE/SEI 12.8.3

Because $0.5 \ s < T = 0.77 \ s < 2.5 \ s$, $k = 0.75 + 0.5T = 1.14$.

Step 14—Determine the lateral seismic force, F_x, at each level   ASCE/SEI 12.8.3

The force F_x is determined by ASCE/SEI Equations (12.8-11) and (12.8-12). A summary of the lateral seismic forces, F_x, and the story shears, V_x, are given in Table 3.28.

Level	Story Weight, w_x (kips)	Height, h_x (ft)	$w_x h_x^k$	Lateral Force, F_x (kips)	Story Shear, V_x (kips)
R	708	110.0	150,391	352.9	352.9
10	812	99.0	152,961	359.0	711.9
9	812	88.0	133,742	313.9	1,025.8
8	812	77.0	114,856	269.6	1,295.4
7	812	66.0	96,347	226.1	1,521.5
6	812	55.0	78,265	183.7	1,705.2
5	812	44.0	60,687	142.4	1,847.6
4	812	33.0	43,718	102.6	1,950.2
3	812	22.0	27,537	64.6	2,014.8
2	812	11.0	12,495	29.3	2,044.1
Σ	8,016		870,999	2,044.1	

1 ft = 0.3048 m; 1 kip = 4.4482 kN.

TABLE 3.28 Seismic Forces and Seismic Story Shears for the Residential Building in Example 3.7

For example, at the second level:

$$C_{vx} = \frac{w_x h_x^k}{\sum w_i h_i} = \frac{812 \times (11.0)^{1.14}}{870,999} = 0.0143$$

$$F_x = C_{vx} V = 0.0143 \times 2,044.1 = 29.3 \text{ kips}$$

In S.I.:

$$C_{vx} = \frac{3,611.9 \times (3.35)^{1.14}}{1,002,192} = 0.0143$$

$$F_x = 0.0143 \times 9,092.5 = 130.0 \text{ kN}$$

3.16.8 Example 3.8—Story Drift Determination

Determine the story drifts for seismic forces in the north-south direction for the 10-story residential building in Fig. 3.35 and check the allowable story drift requirements.

Solution

Step 1—Determine the story drifts, Δ   ASCE/SEI 12.8.6

To check drift limits, the deflections are determined by ASCE/SEI Equation (12.8-15):

$$\delta_x = \frac{C_d \delta_{xe}}{I_e}$$

For a building frame system with special reinforced concrete shear walls, the deflection amplification factor, C_d, is equal to 5 (see ASCE/SEI Table 12.2-1). The displacements, δ_{xe}, at the CM due to the seismic forces F_x are obtained from the three-dimensional analysis and are given in Table 3.29. Also given in Table 3.29 are the interstory drifts, Δ, which are equal to the displacements at the top of the story minus the displacements at the bottom of the story, that is, $\Delta = \delta_x - \delta_{x-1}$. The seismic importance factor, I_e, is equal to 1.0 for Risk Category II buildings.

Step 2—Determine the allowable story drifts, Δ_a   ASCE/SEI 12.12.1

The design story drifts, Δ, must be less than or equal to the allowable story drift, Δ_a, given in ASCE/SEI Table 12.12-1. For Risk Category II and "all other structures," $\Delta_a = 0.020 h_{sx} = 0.020 \times 11.0 \times 12 = 2.64$ in. (67.1 mm) for all stories in this building.

It is evident from Table 3.29 that $\Delta < \Delta_a$ in all stories, so the drift limits are satisfied in the north-south direction.

3.16.9 Example 3.9—P-delta Effects

Determine the P-delta effects for seismic forces in the north-south direction for the 10-story residential building in Fig. 3.35. Assume the live load on the roof is 20 lb/ft² (0.96 kN/m²) and is 40 lb/ft² (1.92 kN/m²) on a typical floor.

Story	δ_{xe}, in. (mm)	δ_x, in. (mm)	Δ, in. (mm)
10	3.08 (78.2)	15.40 (391.0)	2.05 (52.0)
9	2.67 (67.8)	13.35 (339.0)	2.05 (52.0)
8	2.26 (57.4)	11.30 (287.0)	2.05 (52.0)
7	1.85 (47.0)	9.25 (235.0)	1.95 (49.5)
6	1.46 (37.1)	7.30 (185.5)	1.85 (47.0)
5	1.09 (27.7)	5.45 (138.5)	1.70 (43.0)
4	0.75 (19.1)	3.75 (95.5)	1.50 (38.5)
3	0.45 (11.4)	2.25 (57.0)	1.15 (29.0)
2	0.22 (5.6)	1.10 (28.0)	0.75 (19.0)
1	0.07 (1.8)	0.35 (9.0)	0.35 (9.0)

TABLE 3.29 Horizontal Displacements and Story Drifts in the North-South Direction for the Residential Building in Example 3.8

Solution

Step 1—Determine the P-delta effects   ASCE/SEI 12.8.7

In lieu of automatically considering P-delta effects in the computer program analysis, the following procedure can be used to determine if P-delta effects need to be considered.

The stability coefficient, θ, is determined by ASCE/SEI Equation (12.8-16):

$$\theta = \frac{P_x \Delta I_e}{V_x h_{sx} C_d}$$

In this equation, P_x is the total unfactored vertical design load (that is, the total unfactored dead and reduced live loads) at and above level x, V_x is the seismic shear force acting between levels x and $x - 1$, and h_{sx} is the story height below level x.

P-delta calculations for the north-south direction are given in Table 3.30. The floor live load is reduced in accordance with ASCE/SEI 4.7.2. Lateral forces, V_x, are given in Table 3.28, story drifts, Δ, are given in Table 3.29, and $C_d = 5$ from ASCE/SEI Table 12.2-1.

Level	h_{sx}, ft (m)	P_x, kips (kN)	V_x, kips (kN)	Δ, in. (mm)	θ
R	11.0 (3.4)	820.5 (3,649.8)	352.9 (1,569.8)	2.05 (52.0)	0.0072
10	11.0 (3.4)	1,722.5 (7,662.0)	711.9 (3,166.7)	2.05 (52.0)	0.0075
9	11.0 (3.4)	2,624.5 (11,674.3)	1,025.8 (4,563.0)	2.05 (52.0)	0.0079
8	11.0 (3.4)	3,526.5 (15,686.6)	1,295.4 (5,762.2)	1.95 (49.5)	0.0080
7	11.0 (3.4)	4,428.5 (19,698.9)	1,521.5 (6,767.9)	1.85 (47.0)	0.0082
6	11.0 (3.4)	5,330.5 (23,711.1)	1,705.2 (7,585.1)	1.70 (43.0)	0.0081
5	11.0 (3.4)	6,232.5 (27,723.4)	1,847.6 (8,218.5)	1.50 (38.5)	0.0077
4	11.0 (3.4)	7,134.5 (31,735.7)	1,950.2 (8,674.9)	1.15 (29.0)	0.0064
3	11.0 (3.4)	8,036.5 (35,748.0)	2,014.8 (8,962.2)	0.75 (19.0)	0.0045
2	11.0 (3.4)	8,938.5 (39,760.2)	2,044.1 (9,092.6)	0.35 (9.0)	0.0023

TABLE 3.30 P-Delta Effects in the North-South Direction for the Residential Building in Example 3.9

Step 2—Determine if P-delta effects must be considered   ASCE/SEI 12.8.7

P-delta effects need not be considered where the stability coefficient, θ, determined by ASCE/SEI Equation (12.8-16) is less than or equal to 0.10.

It is evident from Table 3.30 that P-delta effects need not be considered at any of the levels because $\theta < 0.10$. Also, $\theta < \theta_{max}$ where θ_{max} is determined by ASCE/SEI Equation (12.8-17):

$$\theta_{max} = \frac{0.5}{\beta C_d} = \frac{0.5}{1.0 \times 5} = 0.1000 < 0.25$$

where the ratio of shear demand to shear capacity, β, is conservatively taken equal to 1.0 in accordance with ASCE/SEI 12.8.7.

3.16.10   Example 3.10—Diaphragm Design Forces (ASCE/SEI 12.10.1)

Determine the diaphragm design forces in the north-south direction for the 10-story residential building in Fig. 3.35 using the requirements in ASCE/SEI 12.10.1. Also, $S_{DS} = 1.561$ from Example 3.7.

Solution

The diaphragm design force, F_{px}, at level x is determined by ASCE/SEI Equation (12.10-1), including the minimum and maximum limits in ASCE/SEI Equations (12.10-2) and (12.10-3), respectively:

$$F_{px} = \frac{\displaystyle\sum_{i=x}^{n} F_i}{\displaystyle\sum_{i=x}^{n} w_i} w_{px}$$

$$\geq 0.2 S_{DS} I_e w_{px} = 0.3122 w_{px}$$

$$\leq 0.4 S_{DS} I_e w_{px} = 0.6244 w_{px}$$

In these equations, w_i is the weight tributary to level i and w_{px} is the weight tributary to the diaphragm at level x. For structures with walls like the building in this example, the weights of the walls parallel to the direction of analysis are often not included in w_{px} because these weights do not contribute to the diaphragm shear forces. However, including the weights of these walls in w_{px} is conservative. In this example, $w_{px} = w_x$ at level x.

A summary of the design diaphragm forces in the north-south direction is given in Table 3.31.

For example, at level 2:

$$F_{p2} = \frac{2,044.1}{8,016} \times 812 = 207.1 \text{ kips } (921.1 \text{ kN})$$

$$\text{Minimum } F_{p2} = 0.2 \times 1.561 \times 1.0 \times 812 = 253.5 \text{ kips } (1,127.7 \text{ kN})$$

Level	w_i (kips)	w_{px} (kips)	F_i (kips)	Σw_i (kips)	ΣF_i (kips)	F_{px} (kips)	Min. F_{px} (kips)	Max. F_{px} (kips)	Design Force (kips)
R	708	708	352.9	708	352.9	352.9	221.0	442.1	352.9
10	812	812	359.0	1,520	711.9	380.3	253.5	507.0	380.3
9	812	812	313.9	2,332	1,025.8	357.2	253.5	507.0	357.2
8	812	812	269.6	3,144	1,295.4	334.6	253.5	507.0	334.6
7	812	812	226.1	3,956	1,521.5	312.3	253.5	507.0	312.3
6	812	812	183.7	4,768	1,705.2	290.4	253.5	507.0	290.4
5	812	812	142.4	5,580	1,847.6	268.9	253.5	507.0	268.9
4	812	812	102.6	6,392	1,950.2	247.7	253.5	507.0	253.5
3	812	812	64.6	7,204	2,014.8	227.1	253.5	507.0	253.5
2	812	812	29.3	8,016	2,044.1	207.1	253.5	507.0	253.5

1 kip = 4.4482 kN.

TABLE 3.31   Design Diaphragm Forces for the Residential Building in Example 3.10

$$\text{Maximum } F_{p2} = 0.4 \times 1.561 \times 1.0 \times 812 = 507.0 \text{ kips } (2,255.3 \text{ kN})$$

Diaphragm design force

$$= \text{maximum} \begin{cases} \text{maximum} \begin{cases} F_2 = 29.3 \text{ kips } (130.3 \text{ kN}) \\ F_{p2} = 207.1 \text{ kips } (921.1 \text{ kN}) \end{cases} \\ \text{minimum } F_{p2} = 253.5 \text{ kips } (1,127.7 \text{ kN}) \text{ (governs)} \end{cases}$$

3.16.11 Example 3.11—Diaphragm Design Forces (ASCE/SEI 12.10.3)

Determine the diaphragm design forces in the north-south direction for the 10-story residential building in Fig. 3.35 using the requirements in ASCE/SEI 12.10.3.

Solution

The flowchart in Fig. 3.24 is used to determine F_{px} at each level.

Step 1—Determine the mode shape factor, z_s   ASCE/SEI 12.10.3.2

For buildings with "other SFRSs," $z_s = 1.0$.

Step 2—Determine the modal contribution factors, Γ_{m1} and Γ_{m2}

ASCE/SEI 12.10.3.2

$$\Gamma_{m1} = 1 + \frac{z_s}{2}\left(1 - \frac{1}{N}\right) = 1 + \left[\frac{1.0}{2} \times \left(1 - \frac{1}{10}\right)\right] = 1.45$$

$$\Gamma_{m2} = 0.9 z_s \left(1 - \frac{1}{N}\right)^2 = 0.9 \times 1.0 \times \left(1 - \frac{1}{10}\right)^2 = 0.73$$

Step 3—Determine the seismic importance factor, I_e   ASCE/SEI Table 1.5-2

For Risk Category II buildings, $I_e = 1.00$.

Step 4—Determine S_{DS} and S_{D1}

From step 4 in Example 3.7, $S_{DS} = 1.561$ and $S_{D1} = 0.787$.

Step 5—Determine the higher mode seismic response coefficient, C_{s2}

Because $N = 10 > 2$, C_{s2} is determined by ASCE/SEI Equations (12.10-10), (12.10-11), and (12.10-12a):

$$C_{s2} = \text{smallest of} \begin{cases} (0.15N + 0.25)I_e S_{DS} = [(0.15 \times 10) + 0.25] \times 1.0 \times 1.561 = 2.732 \\ I_e S_{DS} = 1.0 \times 1.561 = 1.561 \\ \dfrac{I_e S_{D1}}{0.03(N-1)} = \dfrac{1.0 \times 0.787}{0.03 \times (10-1)} = 2.915 \end{cases}$$

Step 6—Determine the design acceleration coefficient, C_{p0}

ASCE/SEI Equation (12.10-6)

$$C_{p0} = 0.4 S_{DS} I_e = 0.4 \times 1.561 \times 1.0 = 0.624$$

Step 7—Determine the overstrength factor, Ω_0   ASCE/SEI Table 12.2-1

For a building frame system with special reinforced concrete shear walls, $\Omega_0 = 2.5$.

Step 8—Determine the seismic response coefficient, C_s

From step 10 in Example 3.7, $C_s = 0.255$.

Step 9—Determine the design acceleration coefficient, C_{pi}, at 80 percent of the structure height

$$C_{pi} = \text{greater of} \begin{cases} C_{p0} = 0.624 \\ 0.9\Gamma_{m1}\Omega_0 C_s = 0.9 \times 1.45 \times 2.5 \times 0.255 = 0.832 \end{cases}$$

Step 10—Determine the design acceleration coefficient, C_{pn}, at the structure height, h_n

$$C_{pn} = \sqrt{(\Gamma_{m1}\Omega_0 C_s)^2 + (\Gamma_{m2}C_{s2})^2}$$
$$= \sqrt{(1.45 \times 2.5 \times 0.255)^2 + (0.73 \times 1.561)^2} = 1.467 > C_{pi} = 0.832$$

Step 11—Determine the reduction factor, R_s   ASCE/SEI Table 12.10-1

Cast-in-place reinforced concrete slabs designed in accordance and ACI 318 are typically controlled by in-plane flexure. Thus, $R_s = 2.0$.

Step 12—Determine the design acceleration coefficient, C_{px}, at level x
         ASCE/SEI Figure 12.10-2

For $h_x \leq 0.8h_n$:

$$C_{px} = C_{p0} + \left(\frac{C_{pi} - C_{p0}}{0.8}\right)\left(\frac{h_x}{h_n}\right) = 0.624 + \left[0.260 \times \left(\frac{h_x}{110.0}\right)\right]$$

For $h_x > 0.8h_n$:

$$C_{px} = C_{pi} + \left(\frac{C_{pn} - C_{pi}}{0.2}\right)\left(\frac{h_x}{h_n} - 0.8\right) = 0.832 + \left[3.175 \times \left(\frac{h_x}{110.0} - 0.8\right)\right]$$

Step 13—Determine the diaphragm design force, F_{px}, at level x

The diaphragm design force, F_{px}, at level x is determined by ASCE/SEI Equation (12.10-4), including the minimum limit in ASCE/SEI Equation (12.10-5):

$$F_{px} = \frac{C_{px}}{R_s} w_{px} \geq 0.2 S_{DS} I_e w_{px} = 0.3122 w_{px}$$

A summary of the design diaphragm forces in the north-south direction is given in Table 3.32.

For example, at the third level:

$$h_x = 22.0 \text{ ft } (6.7 \text{ m}) < 0.8h_n = 0.8 \times 110.0 = 88.0 \text{ ft } (26.8 \text{ m})$$

$$C_{px} = 0.624 + \left[0.260 \times \left(\frac{22.0}{110}\right)\right] = 0.676$$

Level	W_{px} (kips)	Height, h_x (ft)	h_x/h_n	C_{px}	F_{px} (kips)	Min. F_{px} (kips)	Design Force (kips)
R	708	110.0	1.0	1.467	519.3	221.0	519.3
10	812	99.0	0.9	1.150	466.9	253.5	466.9
9	812	88.0	0.8	0.832	337.8	253.5	337.8
8	812	77.0	0.7	0.806	327.2	253.5	327.2
7	812	66.0	0.6	0.780	316.7	253.5	316.7
6	812	55.0	0.5	0.754	306.1	253.5	306.1
5	812	44.0	0.4	0.728	295.6	253.5	295.6
4	812	33.0	0.3	0.702	285.0	253.5	285.0
3	812	22.0	0.2	0.676	274.5	253.5	274.5
2	812	11.0	0.1	0.650	263.9	253.5	263.9

1 ft = 0.3048 m
1 kip = 4.4482 kN.

TABLE 3.32 Design Diaphragm Forces for the Residential Building in Example 3.11

$$F_{px} = \frac{C_{px}}{R_s} w_{px} = \frac{0.676}{2.0} \times 812 = 274.5 \text{ kips } (1,222.3 \text{ kN})$$
$$\geq 0.2 S_{DS} I_e w_{px} = 0.3122 \times 812 = 253.5 \text{ kips } (1,127.7 \text{ kN})$$

3.16.12 Example 3.12—Horizontal Distribution of Seismic Forces—Rigid Diaphragm

Determine the seismic forces in the elements of the SFRS in the north-south direction for the one-story commercial building in Fig. 3.36 assuming the diaphragm is rigid. The height of the building is 15.0 ft (4.6 m). Design data are given in Table 3.33.

Solution

Step 1—Determine V and F_x using the ELF procedure   ASCE/SEI 12.8

The flowchart in Fig. 3.20 is used to determine V and F_x. A summary of the calculations is given in Table 3.34.

Step 2—Determine the diaphragm force, F_{px}, at the roof level   ASCE/SEI 12.10.1.1

$$F_{px} = F_{pR} = \frac{F_R}{w_R} w_{pR} = F_R = 58.8 \text{ kips } (262.0 \text{ kN})$$

Minimum $F_{pR} = 0.2 S_{DS} I_e w_{pR} = 0.2 \times 0.621 \times 1.0 \times 565 = 70.2$ kips

Maximum $F_{pR} = 0.4 S_{DS} I_e w_{pR} = 0.4 \times 0.621 \times 1.0 \times 565 = 140.4$ kips

In S.I.:

Minimum $F_{pR} = 0.2 \times 0.621 \times 1.0 \times 2,519 = 312.9$ kN

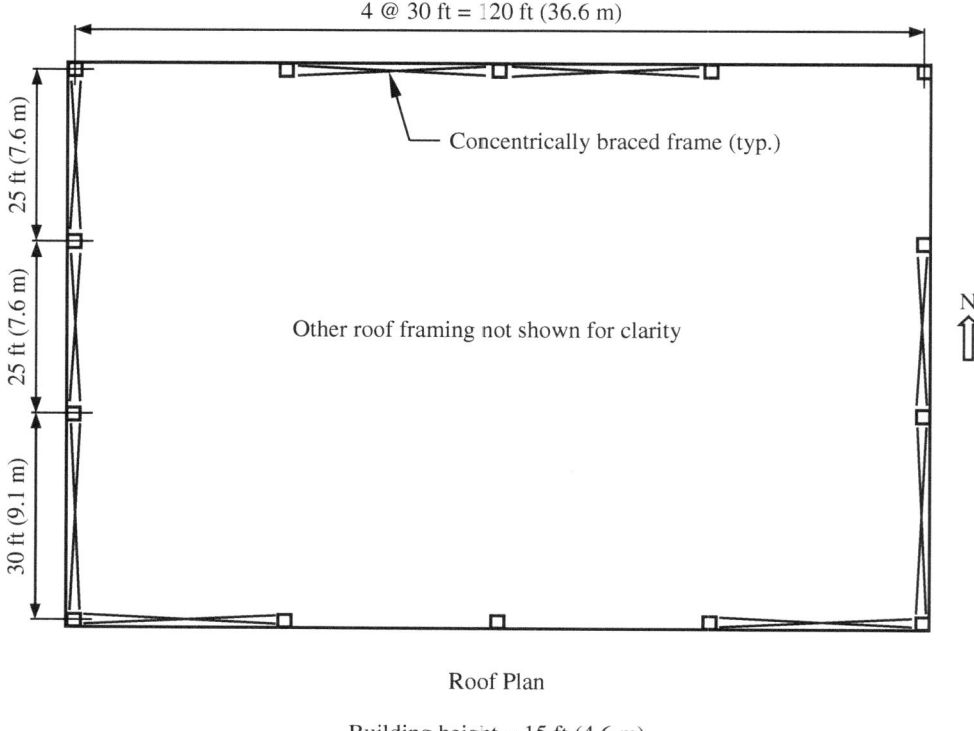

4 @ 30 ft = 120 ft (36.6 m)

25 ft (7.6 m)

25 ft (7.6 m)

30 ft (9.1 m)

Concentrically braced frame (typ.)

Other roof framing not shown for clarity

N

Roof Plan

Building height = 15 ft (4.6 m)

Figure 3.36 Roof plan of the commercial building in Example 3.12.

Maximum $F_{pR} = 0.4 \times 0.621 \times 1.0 \times 2,519 = 625.8$ kN

Therefore, $F_{pR} = 70.2$ kips (312.9 kN).

Step 3—Determine the relative stiffnesses of the elements of the SFRS

Approximate relative stiffnesses of the concentrically braced frames are determined using Fig. 3.14. Assuming a double diagonal configuration where

Location	Latitude: 37.726°, Longitude: −89.217°
Soil classification	Site class C
Occupancy	Commercial occupancy where less than 300 people congregate in one area
SFRS	Building frame with steel special concentrically braced frames (system B2)
Weight of roof framing and superimposed dead load	54 lb/ft² (2.59 kN/m²)
Weight of cladding	8 lb/ft² (0.383 kN/m²)

Table 3.33 Design Data for the Commercial Building in Example 3.12

$S_S = 0.777$ $S_1 = 0.270$	Refs. 5 through 7
$F_a = 1.20$ $F_v = 1.50$	ASCE/SEI Tables 11.4-1 and 11.4-2
$S_{MS} = F_a S_S = 1.20 \times 0.777 = 0.932$ $S_{M1} = F_v S_1 = 1.50 \times 0.270 = 0.405$	ASCE/SEI Equations (11.4-1) and (11.4-2)
$S_{DS} = 2S_{MS}/3 = 0.621$ $S_{D1} = 2S_{M1}/3 = 0.270$	ASCE/SEI Equations (11.4-3) and (11.4-4)
Risk Category is II for commercial occupancy	ASCE/SEI Table 1.5-1
For Risk Category II and $S_{DS} = 0.621 > 0.50$, SDC is D For Risk Category II and $S_{D1} = 0.270 > 0.20$, SDC is D Therefore, the SDC is D	ASCE/SEI Tables 11.6-1 and 11.6-2
$R = 6$	ASCE/SEI Table 12.2-1
For Risk Category II, $I_e = 1.0$	ASCE/SEI Table 1.5-2
$T_a = C_t h_n^x$ For "other structural systems": $C_t = 0.02$ $x = 0.75$ $T_a = 0.02 \times 15.0^{0.75} = 0.15$ s In S.I.: $C_t = 0.0488$ $x = 0.75$ $T_a = 0.0488 \times 4.6^{0.75} = 0.15$ s	ASCE/SEI 12.8.2.1
$T_L = 12$ s	Refs. 5 through 7
$C_s = \dfrac{S_{DS}}{\left(\dfrac{R}{I_e}\right)} = \dfrac{0.621}{\left(\dfrac{6}{1.0}\right)} = 0.104$ $\leq \dfrac{S_{D1}}{T\left(\dfrac{R}{I_e}\right)} = \dfrac{0.270}{0.15 \times \left(\dfrac{6}{1.0}\right)} = 0.300$ $\geq$ larger of $\begin{cases} 0.044 S_{DS} I_e = 0.027 \\ 0.01 \end{cases}$ Therefore, $C_s = 0.104$	ASCE/SEI Equations (12.8-2), (12.8-3), and (12.8-5)
Weight of roof framing and superimposed dead load $= 0.054 \times 82.0 \times 122.0 = 540$ kips Weight of cladding $= 0.008 \times (15.0/2) \times [2 \times (122.0 + 82.0)] = 25$ kips $W = 540 + 25 = 565$ kips	ASCE/SEI 12.7.2

TABLE 3.34 Summary of Force Calculations for the Commercial Building in Example 3.12

In S.I.:	
Weight of roof framing and superimposed dead load $= 2.59 \times 25.0 \times 37.2 = 2,409$ kN	
Weight of cladding $= 0.383 \times (4.6/2) \times [2 \times (25.0 + 37.2)] = 110$ kN	
$W = 2,409 + 110 = 2,519$ kN	
$V = C_s W = 0.104 \times 565 = 58.8$ kips	ASCE/SEI Equation (12.8-1)
In S.I.:	
$V = 0.104 \times 2,519 = 262.0$ kN	
For $T = 0.15$ s < 0.5 s, $k = 1.0$	ASCE/SEI 12.8.3
$F_x = F_R = V = 58.8$ kips (262.0 kN)	ASCE/SEI Equations (12.8-11) and (12.8-12)

TABLE 3.34   Summary of Force Calculations for the Commercial Building in Example 3.12 (*Continued*)

the sizes of the members are the same for each frame, the stiffness, k_i, can be determined from the following equation (the area of the braces, A_{br}, and the modulus of elasticity, E, are the same for all frames, so both cancel out from relative stiffness calculations):

$$k_i = \frac{2E}{\left(\dfrac{\ell_{br}^3}{A_{br}\ell_i^2}\right)} = \frac{2\ell_i^2}{\ell_{br}^3}$$

For the 25-ft frames:

$$\ell_i = 25.0 \text{ ft}$$

$$\ell_{br} = \sqrt{25.0^2 + 15.0^2} = 29.2 \text{ ft}$$

$$k_i = \frac{2 \times 25.0^2}{29.2^3} = 0.0502$$

For the 30-ft frames:

$$\ell_i = 30.0 \text{ ft}$$

$$\ell_{br} = \sqrt{30.0^2 + 15.0^2} = 33.5 \text{ ft}$$

$$k_i = \frac{2 \times 30.0^2}{33.5^3} = 0.0479$$

In S.I.:

For the 7.6-m frames:

$$\ell_i = 7.6 \text{ m}$$

$$\ell_{br} = \sqrt{7.6^2 + 4.6^2} = 8.9 \text{ m}$$

$$k_i = \frac{2 \times 7.6^2}{8.9^3} = 0.1639$$

For the 9.1-m frames:

$$\ell_i = 9.1 \text{ m}$$

$$\ell_{br} = \sqrt{9.1^2 + 4.6^2} = 10.2 \text{ m}$$

$$k_i = \frac{2 \times 9.1^2}{10.2^3} = 0.1561$$

Step 4—Determine the location of the CM

The majority of the roof weight is due to the roof framing. The CM is located from the west and south frame centerlines as follows:

$$x_{cm} = 60.0 \text{ ft } (18.3 \text{ m})$$
$$y_{cm} = 40.0 \text{ ft } (12.2 \text{ m})$$

Step 5—Determine the location of the CR

The CR is determined using the intersection of the centerlines of the west and south frames as the origin. The equations in Fig. 3.13 are used to determine the location of the CR:

$$x_{cr} = \frac{(k_i)_y x_i}{\sum (k_i)_y} = \frac{(0.0502 + 0.0479) \times 120.0}{(3 \times 0.0502) + (2 \times 0.0479)} = 47.8 \text{ ft}$$

$$y_{cr} = \frac{(k_i)_x y_i}{\sum (k_i)_x} = \frac{2 \times 0.0479 \times 80.0}{4 \times 0.0479} = 40.0 \text{ ft}$$

In S.I.:

$$x_{cr} = \frac{(k_i)_y x_i}{\sum (k_i)_y} = \frac{(0.1639 + 0.1561) \times 36.6}{(3 \times 0.1639) + (2 \times 0.1561)} = 14.6 \text{ m}$$

$$y_{cr} = \frac{(k_i)_x y_i}{\sum (k_i)_x} = \frac{2 \times 0.1561 \times 24.4}{4 \times 0.1561} = 12.2 \text{ m}$$

As expected, y_{cr} is located halfway between the north and south frames due to symmetry (see Fig. 3.37).

Step 6—Distribute the diaphragm seismic force to the vertical elements of the SFRS

The diaphragm force F_{pR} acts through the CM. Because the diaphragm is rigid, the distribution of the diaphragm force to the vertical elements of the SFRS must consider the effects of the inherent torsional moment, M_t, due to the eccentricity, e_x, between the locations of the CM and CR.

Force calculations are given in Fig. 3.38 for seismic forces in the north direction using the equations in Fig. 3.15. The forces indicated in the figure are in the directions that resist the in-plane seismic force and inherent torsional moment. Forces indicated by a solid arrow are term 1 forces (forces due to F_{pR}) and those indicated by a dashed arrow are term 2 forces (forces due to M_t).

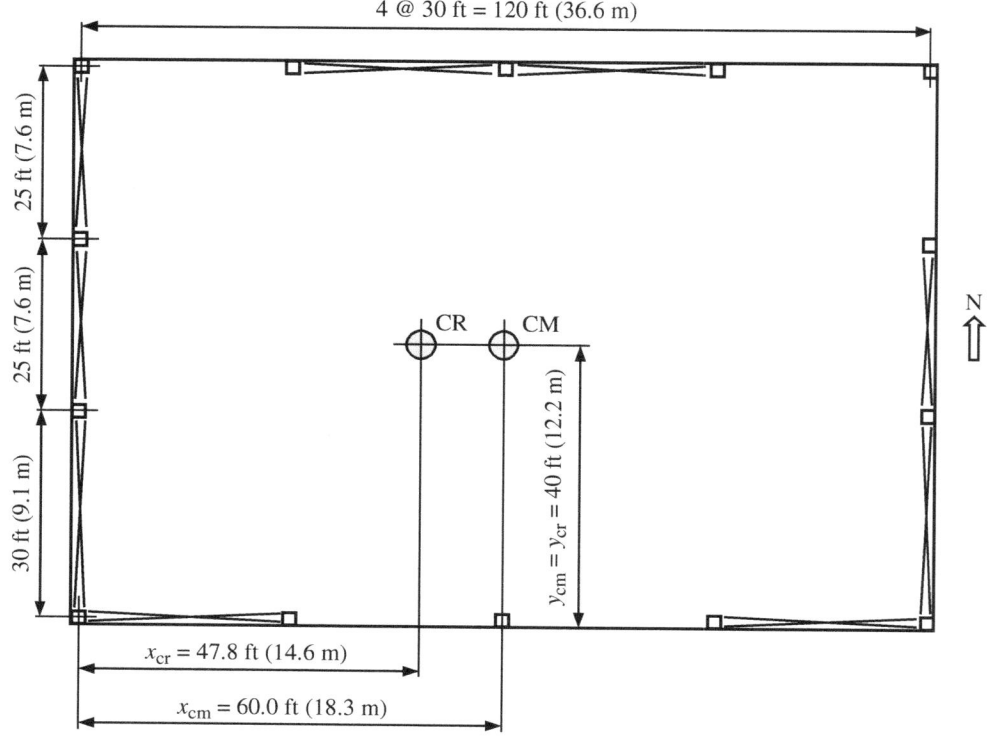

Figure 3.37   Location of the CM and CR for the commercial building in Example 3.12.

For example, the forces in the braced frame labeled 2 are determined as follows:

Term 1 force:

$$(V_2)_1 = \frac{(k_2)_y F_{pr}}{\sum (k_i)_y} = \frac{0.0502 \times 70.2}{0.2464} = 14.30 \text{ kips}$$

Term 2 force:

$$\bar{x}_2^2(k_2)_y = 47.8^2 \times 0.0502 = 114.7$$

$$(V_2)_2 = \frac{-\bar{x}_2(k_2)_y F_{pr} e_x}{\sum \bar{x}_i^2 (k_i)_y + \sum \bar{y}_i^2 (k_i)_x} = \frac{-47.8 \times 0.0502 \times 70.2 \times 12.2}{850.2 + 306.4} = -1.78 \text{ kips}$$

Total force $V_2 = 14.30 - 1.78 = 12.52$ kips

In S.I.:

Term 1 force:

$$(V_2)_1 = \frac{(k_2)_y F_{pr}}{\sum (k_i)_y} = \frac{0.1639 \times 312.9}{0.8039} = 63.79 \text{ kN}$$

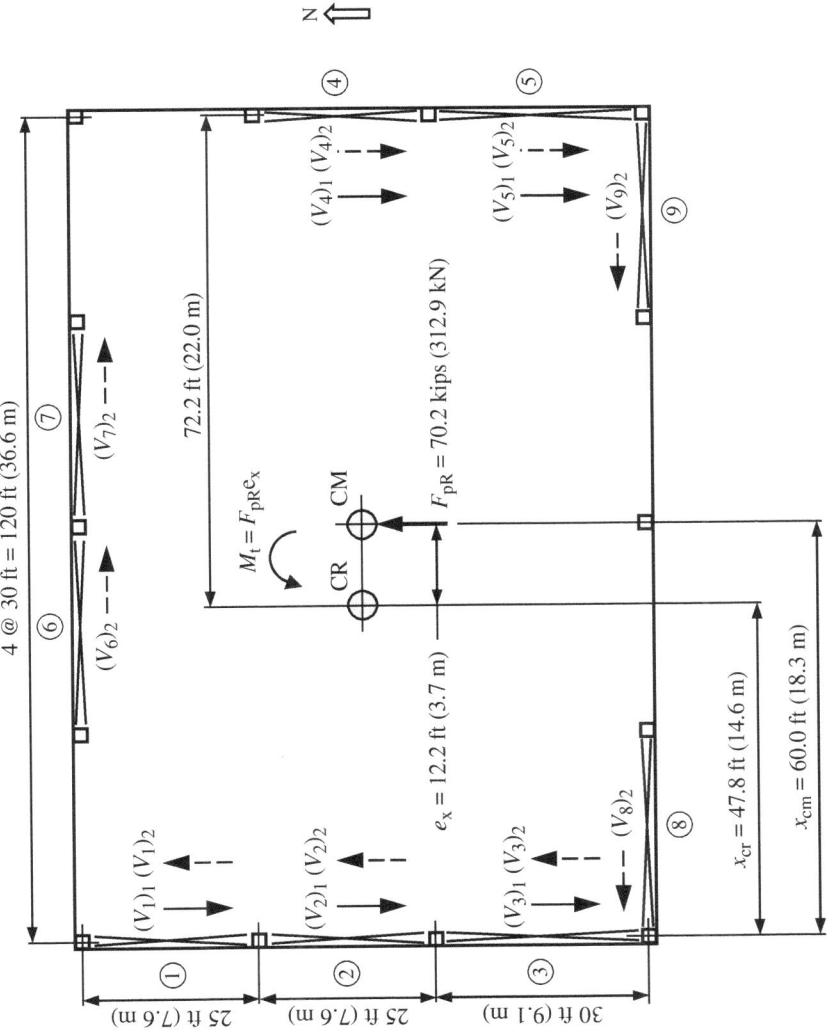

Figure 3.38 Force allocation to the braced frames for the commercial building in Example 3.12.

Frame	x_i (ft)	y_i (ft)	$(k_i)_y$	$(k_i)_x$	$\bar{x}_i$ (ft)	$\bar{y}_i$ (ft)	$\bar{x}_i^2(k_i)_y$ (ft)	$\bar{y}_i^2(k_i)_x$ (ft)	Term 1 $(V_i)_1 = \dfrac{(k_i)_y F_{pr}}{\Sigma(k_i)_y}$ (kips)	Term 2 $(V_i)_2 = \left(\dfrac{\bar{x}_i(k_i)_y F_{pr} e_x}{\Sigma \bar{x}_i^2(k_i)_y + \Sigma \bar{y}_i^2(k_i)_x}\right)^*$ (kips)	$V_i = (V_i)_1 + (V_i)_2$ (kips)
1	0.0	–	0.0502	–	47.8	–	114.7	–	14.30	–1.78	12.52
2	0.0	–	0.0502	–	47.8	–	114.7	–	14.30	–1.78	12.52
3	0.0	–	0.0479	–	47.8	–	109.4	–	13.65	–1.70	11.95
4	120.0	–	0.0502	–	72.2	–	261.7	–	14.30	2.68	16.98
5	120.0	–	0.0479	–	72.2	–	249.7	–	13.65	2.56	16.21
6	–	80.0	–	0.0479	–	40.0	–	76.6	–	1.42	1.42
7	–	80.0	–	0.0479	–	40.0	–	76.6	–	1.42	1.42
8	–	0.0	–	0.0479	–	40.0	–	76.6	–	–1.42	–1.42
9	–	0.0	–	0.0479	–	40.0	–	76.6	–	–1.42	–1.42
Σ			0.2464	0.1916			850.2	306.4			70.18

Frame	x_i (m)	y_i (m)	$(k_i)_y$	$(k_i)_x$	$\bar{x}_i$ (m)	$\bar{y}_i$ (m)	$\bar{x}_i^2(k_i)_y$ (m)	$\bar{y}_i^2(k_i)_x$ (m)	Term 1 $(V_i)_1 = \dfrac{(k_i)_y F_{pr}}{\Sigma(k_i)_y}$ (kN)	Term 2 $(V_i)_2 = \left(\dfrac{\bar{x}_i(k_i)_y F_{pr} e_x}{\Sigma \bar{x}_i^2(k_i)_y + \Sigma \bar{y}_i^2(k_i)_x}\right)^*$ (kN)	$V_i = (V_i)_1 + (V_i)_2$ (kN)
1	0.0	–	0.1639	–	14.6	–	34.94	–	63.79	–7.89	55.90
2	0.0	–	0.1639	–	14.6	–	34.94	–	63.79	–7.89	55.90
3	0.0	–	0.1561	–	14.6	–	33.27	–	60.76	–7.52	53.24
4	36.6	–	0.1639	–	22.0	–	79.33	–	63.79	11.90	75.69
5	36.6	–	0.1561	–	22.0	–	75.55	–	60.76	11.33	72.09
6	–	24.4	–	0.1561	–	12.2	–	23.23	–	6.28	6.28
7	–	24.4	–	0.1561	–	12.2	–	23.23	–	6.28	6.28
8	–	0.0	–	0.1561	–	12.2	–	23.23	–	–6.28	–6.28
9	–	0.0	–	0.1561	–	12.2	–	23.23	–	–6.28	–6.28
Σ			0.8039	0.6244			258.03	92.92			312.82

*For frames 7, 8, and 9, replace $\bar{x}_i(k_i)_y$ with $\bar{y}_i(k_i)_x$.

FIGURE 3.38 (Continued)

Term 2 force:

$$\bar{x}_2^2(k_2)_y = 14.6^2 \times 0.1639 = 34.94$$

$$(V_2)_2 = \frac{-\bar{x}_2(k_2)_y F_{pr}e_x}{\sum \bar{x}_i^2(k_i)_y + \sum \bar{y}_i^2(k_i)_x} = \frac{-14.6 \times 0.1639 \times 312.9 \times 3.7}{258.03 + 92.92} = -7.89 \text{ kN}$$

Total force $V_2 = 63.79 - 7.89 = 55.90$ kN

The calculations were performed with two decimal places to demonstrate that the summation of the V_i must equal F_{pr} (considering roundoff).

The term 2 force is negative because it acts in the opposite direction of the term 1 force for F_{pr} acting in the north direction. However, the diaphragm must also be designed for F_{pr} acting in the south direction, so the maximum force that braced frame 2 is subjected to is $14.30 + 1.78 = 16.08$ kips (63.79 + 7.89 = 71.68 kN).

3.16.13   Example 3.13—Horizontal Distribution of Seismic Forces—Flexible Diaphragm

Determine the seismic forces in the elements of the SFRS in the north-south direction for the one-story commercial building in Fig. 3.39 assuming the diaphragm is flexible. The height of the building is 15.0 ft (4.6 m) and there is a 2.0-ft (0.61-m) tall parapet. Design data are given in Table 3.35.

Solution

Step 1—*Determine V and F_x using the ELF procedure*   ASCE/SEI 12.8

The flowchart in Fig. 3.20 is used to determine V and F_x. A summary of the calculations is given in Table 3.36.

Step 2—*Determine the diaphragm force, F_{px}, at the roof level*   ASCE/SEI 12.10.1.1

The diaphragm force is determined considering the weight of the roof framing, the superimposed dead load, and the weight of the masonry walls perpendicular to the direction of analysis. Therefore,

$$w_{pR} = 91 + 74 = 165 \text{ kips}$$

$$F_{px} = F_{pR} = \frac{F_R}{w_R} w_{pR} = \frac{27.5}{243} \times 165 = 18.7 \text{ kips}$$

Minimum $F_{pR} = 0.2S_{DS}I_e w_{pR} = 0.2 \times 0.621 \times 1.0 \times 165 = 20.5$ kips

Maximum $F_{pR} = 0.4S_{DS}I_e w_{pR} = 0.4 \times 0.621 \times 1.0 \times 165 = 41.0$ kips

Therefore, $F_{pR} = F_R = 27.5$ kips > minimum $F_{pR} = 20.5$ kips

In S.I.:

$$w_{pR} = 406 + 329 = 735 \text{ kN}$$

$$F_{px} = F_{pR} = \frac{F_R}{w_R} w_{pR} = \frac{122.4}{1,083} \times 735 = 83.1 \text{ kN}$$

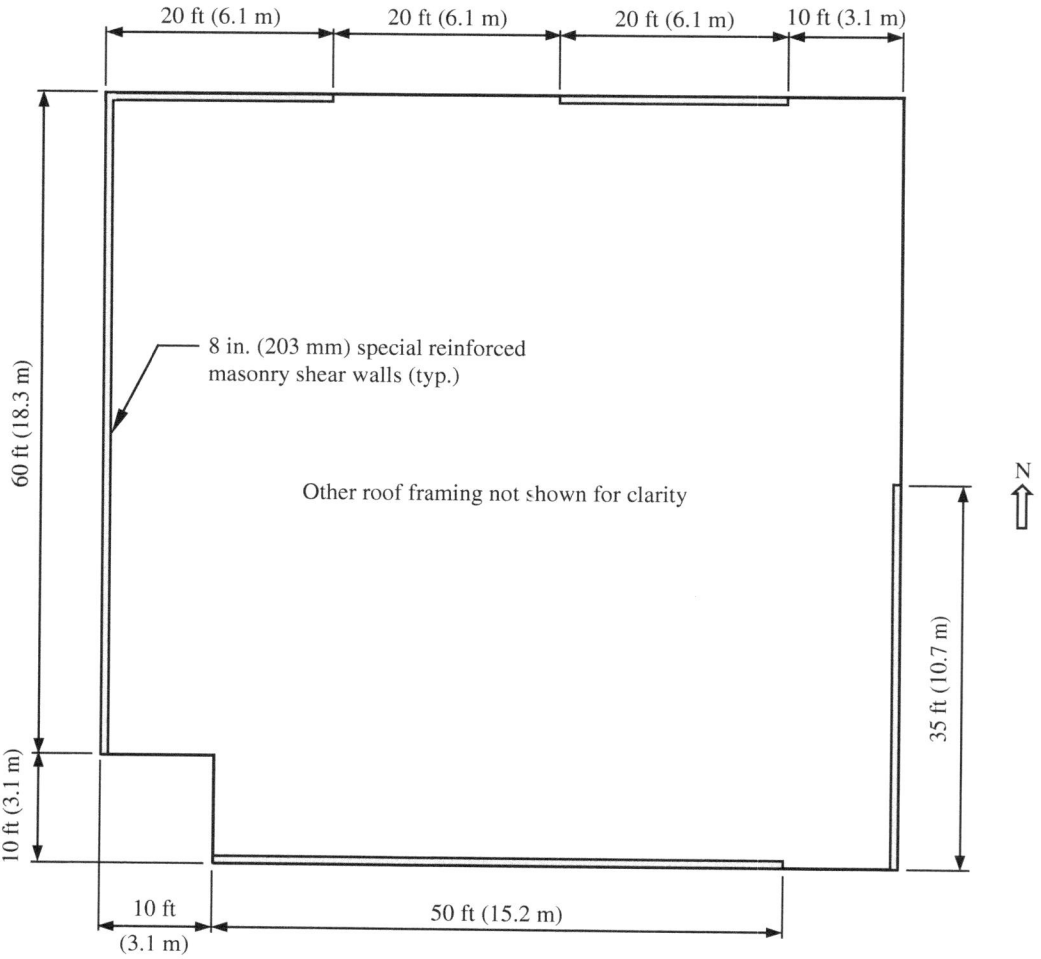

FIGURE 3.39 Roof plan of the commercial building in Example 3.13.

Location	Latitude: 37.726°, Longitude: −89.217°
Soil classification	Site class C
Occupancy	Commercial occupancy where less than 300 people congregate in one area
SFRS	Special reinforced masonry shear walls (system B16)
Weight of roof framing and superimposed dead load	19 lb/ft² (0.91 kN/m²)
Weight of 8 in. (203 mm) concrete masonry walls, fully grouted	86 lb/ft² (4.12 kN/m²)

TABLE 3.35 Design Data for the Commercial Building in Example 3.13

$S_S = 0.777$ $S_1 = 0.270$	Refs. 5 through 7
$F_a = 1.20$ $F_v = 1.50$	ASCE/SEI Tables 11.4-1 and 11.4-2
$S_{MS} = F_a S_S = 1.20 \times 0.777 = 0.932$ $S_{M1} = F_v S_1 = 1.50 \times 0.270 = 0.405$	ASCE/SEI Equations (11.4-1) and (11.4-2)
$S_{DS} = 2S_{MS}/3 = 0.621$ $S_{D1} = 2S_{M1}/3 = 0.270$	ASCE/SEI Equations (11.4-3) and (11.4-4)
Risk Category is II for commercial occupancy	ASCE/SEI Table 1.5-1
For Risk Category II and $S_{DS} = 0.621 > 0.50$, SDC is D For Risk Category II and $S_{D1} = 0.270 > 0.20$, SDC is D Therefore, the SDC is D	ASCE/SEI Tables 11.6-1 and 11.6-2
$R = 5.5$	ASCE/SEI Table 12.2-1
For Risk Category II, $I_e = 1.0$	ASCE/SEI Table 1.5-2
$T_a = C_t h_n^x$ For "other structural systems": $C_t = 0.02$ $x = 0.75$ $T_a = 0.02 \times 15.0^{0.75} = 0.15$ s In S.I.: $C_t = 0.0488$ $x = 0.75$ $T_a = 0.0488 \times 4.6^{0.75} = 0.15$ s	ASCE/SEI 12.8.2.1
$T_L = 12$ s	Refs. 5 through 7
$C_s = \dfrac{S_{DS}}{\left(\dfrac{R}{I_e}\right)} = \dfrac{0.621}{\left(\dfrac{5.5}{1.0}\right)} = 0.113$ $\leq \dfrac{S_{D1}}{T\left(\dfrac{R}{I_e}\right)} = \dfrac{0.270}{0.15 \times \left(\dfrac{5.5}{1.0}\right)} = 0.327$ $\geq$ larger of $\begin{cases} 0.044 S_{DS} I_e = 0.027 \\ 0.01 \end{cases}$ Therefore, $C_s = 0.113$	ASCE/SEI Equations (12.8-2), (12.8-3), and (12.8-5)
Area of roof = $(60.0 \times 70.0) + (10.0 \times 60.0) = 4,800$ ft² (445.9 m²) Weight of roof framing and superimposed dead load $= 0.019 \times 4,800 = 91$ kips Weight of walls parallel to the direction of analysis $= 0.086 \times [(15.0/2) + 2.0] \times (60.0 + 35.0) = 78$ kips	ASCE/SEI 12.7.2

TABLE 3.36 Summary of Force Calculations for the Commercial Building in Example 3.13

Weight of walls perpendicular to the direction of analysis $= 0.086 \times [(15.0/2) + 2.0] \times (50.0 + 20.0 + 20.0) = 74$ kips $W = 91 + 78 + 74 = 243$ kips In S.I.: Weight of roof framing and superimposed dead load $= 0.91 \times 445.9 = 406$ kN Weight of walls parallel to the direction of analysis $= 4.12 \times [(4.6/2) + 0.61] \times (18.3 + 10.7) = 348$ kN Weight of walls perpendicular to the direction of analysis $= 4.12 \times [(4.6/2) + 0.61] \times (15.2 + 6.1 + 6.1) = 329$ kN $W = 406 + 348 + 329 = 1{,}083$ kN	
$V = C_s W = 0.113 \times 243 = 27.5$ kips In S.I.: $V = 0.113 \times 1{,}083 = 122.4$ kN	ASCE/SEI Equation (12.8-1)
For $T = 0.15$ s < 0.5 s, $k = 1.0$	ASCE/SEI 12.8.3
$F_x = F_R = V = 27.5$ kips (122.4 kN)	ASCE/SEI Equations (12.8-11) and (12.8-12)

TABLE 3.36 Summary of Force Calculations for the Commercial Building in Example 3.13 (*Continued*)

$$\text{Minimum } F_{pR} = 0.2 \times 0.621 \times 1.0 \times 735 = 91.3 \text{ kN}$$

$$\text{Maximum } F_{pR} = 0.4 \times 0.621 \times 1.0 \times 735 = 182.6 \text{ kN}$$

$$\text{Therefore, } F_{pR} = F_R = 122.4 \; kN > \text{minimum } F_{pR} = 91.3 \text{ kN}$$

Step 3—Distribute the diaphragm seismic force to the vertical elements of the SFRS

The diaphragm force F_{pR} is converted to a uniformly distributed load over the width of the diaphragm perpendicular to the direction of analysis:

$$w_u = \frac{27.5}{70.0} = 0.393 \text{ kips/ft}$$

Because the diaphragm is flexible, the forces in walls 1 and 2 are equal (see Fig. 3.40):

$$V_1 = V_2 = \frac{0.393 \times 70.0}{2} = 13.8 \text{ kips}$$

In S.I.:

$$w_u = \frac{122.4}{21.3} = 5.75 \text{ kN/m}$$

$$V_1 = V_2 = \frac{5.75 \times 21.3}{2} = 61.2 \text{ kN}$$

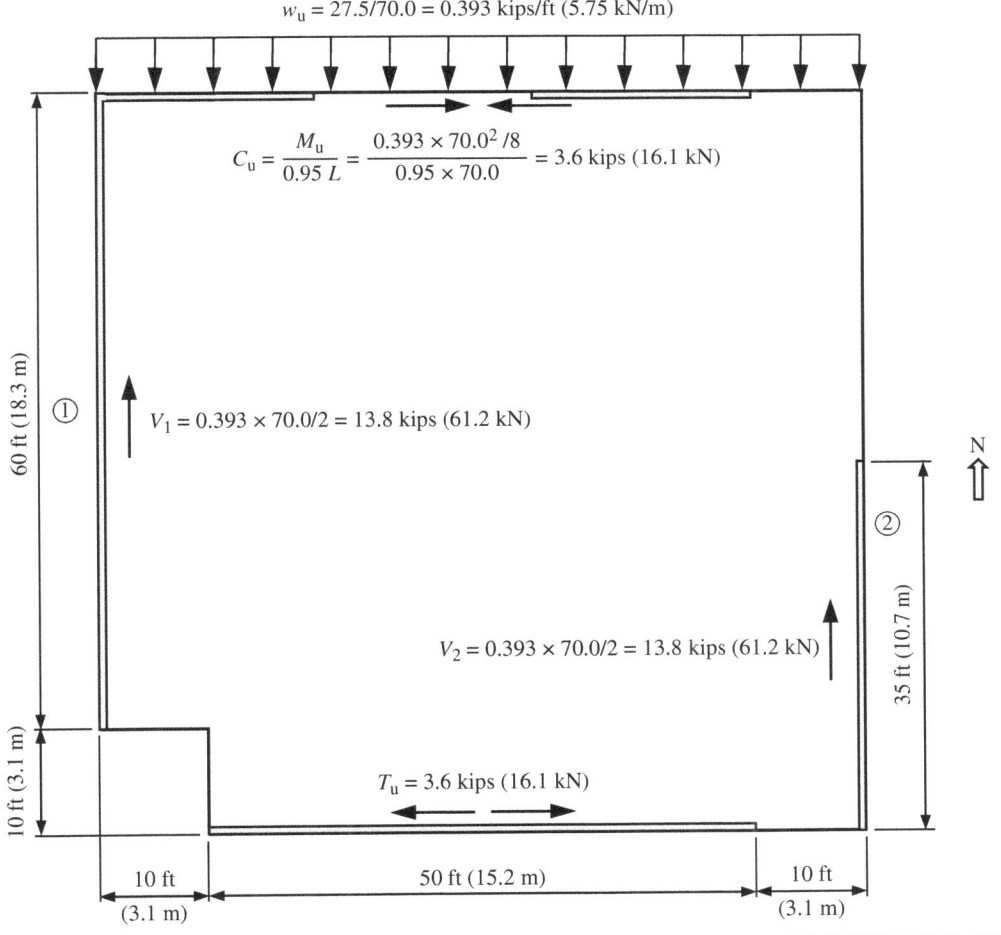

$w_u = 27.5/70.0 = 0.393$ kips/ft (5.75 kN/m)

$$C_u = \frac{M_u}{0.95\,L} = \frac{0.393 \times 70.0^2\,/8}{0.95 \times 70.0} = 3.6 \text{ kips (16.1 kN)}$$

①

$V_1 = 0.393 \times 70.0/2 = 13.8$ kips (61.2 kN)

②

$V_2 = 0.393 \times 70.0/2 = 13.8$ kips (61.2 kN)

$T_u = 3.6$ kips (16.1 kN)

60 ft (18.3 m)

10 ft (3.1 m)

35 ft (10.7 m)

N

10 ft
(3.1 m)

50 ft (15.2 m)

10 ft
(3.1 m)

FIGURE 3.40 Force allocation to the masonry walls for the commercial building in Example 3.13.

Step 4—Determine the chord forces

The compression and tension chord forces, C_u and T_u, are determined based on the maximum in-plane bending moment in the diaphragm (see Fig. 3.40):

$$M_u = \frac{0.393 \times 70.0^2}{8} = 240.7 \text{ ft-kips}$$

$$C_u = T_u = \frac{M_u}{0.95L} = \frac{240.7}{0.95 \times 70.0} = 3.6 \text{ kips}$$

In S.I.:

$$M_u = \frac{5.75 \times 21.3^2}{8} = 326.1 \text{ kN-m}$$

$$C_u = T_u = \frac{326.1}{0.95 \times 21.3} = 16.1 \text{ kN}$$

Step 5—Determine the axial forces in the collector element at wall 2

The unit shear forces, net shear forces, and axial forces at wall 2 are given in Fig. 3.41.

The unit shear force in the diaphragm is determined by dividing the force in the wall, V_2, by the overall depth of the diaphragm in the direction of analysis. Similarly, the unit shear force in wall 2 is determined by dividing V_2 by the length of the wall. The net shear forces are determined by algebraically combining the unit shear forces. The axial forces in the collector are determined by calculating the area under the net shear diagram.

3.16.14 Example 3.14—Structural Wall Out-of-Plane Forces

Determine the out-of-plane seismic force on a masonry wall in Example 3.13.

Solution

Structural walls must be designed for a seismic force, F_p, normal to the surface equal to $0.4S_{DS}I_e$ times the weight of the wall, w_p (ASCE/SEI 12.11.1). The minimum F_p is equal to 10 percent times w_p.

$$\text{Weight of wall } w_p = 86.0 \times 17.0 = 1,462 \text{ lb per foot width of wall}$$

$$F_p = 0.4S_{DS}I_ew_p = 0.4 \times 0.621 \times 1.0 \times 1,462 = 363 \text{ lb per foot width of wall}$$

$$> 0.1w_p = 0.1 \times 1,462 = 146 \text{ lb per foot width of wall}$$

This force acts at the centroid of the wall (see Fig. 3.42).

The out-of-plane seismic force can also be expressed as a uniformly distributed load over the height of the wall:

$$f_p = 363/17.0 = 21.4 \text{ lb/ft per foot width of wall}$$

In S.I.:

$$\text{Weight of wall } w_p = 4.12 \times 5.2 = 21.4 \text{ kN per meter width of wall}$$

$$F_p = 0.4S_{DS}I_ew_p = 0.4 \times 0.621 \times 1.0 \times 21.4 = 5.3 \text{ kN per meter width of wall}$$

$$> 0.1w_p = 0.1 \times 21.4 = 2.1 \text{ kN per meter width of wall}$$

$$f_p = 5.3/5.2 = 1.0 \text{ kN/m per meter width of wall}$$

3.16.15 Example 3.15—Wall Anchorage Forces

Determine the anchorage force between the masonry wall and the roof diaphragm in Example 3.13.

Solution

Wall anchorage force, F_p, is determined by ASCE/SEI Equation (12.11-1):

$$F_p = 0.4S_{DS}k_aI_eW_p \geq 0.2k_aI_eW_p$$

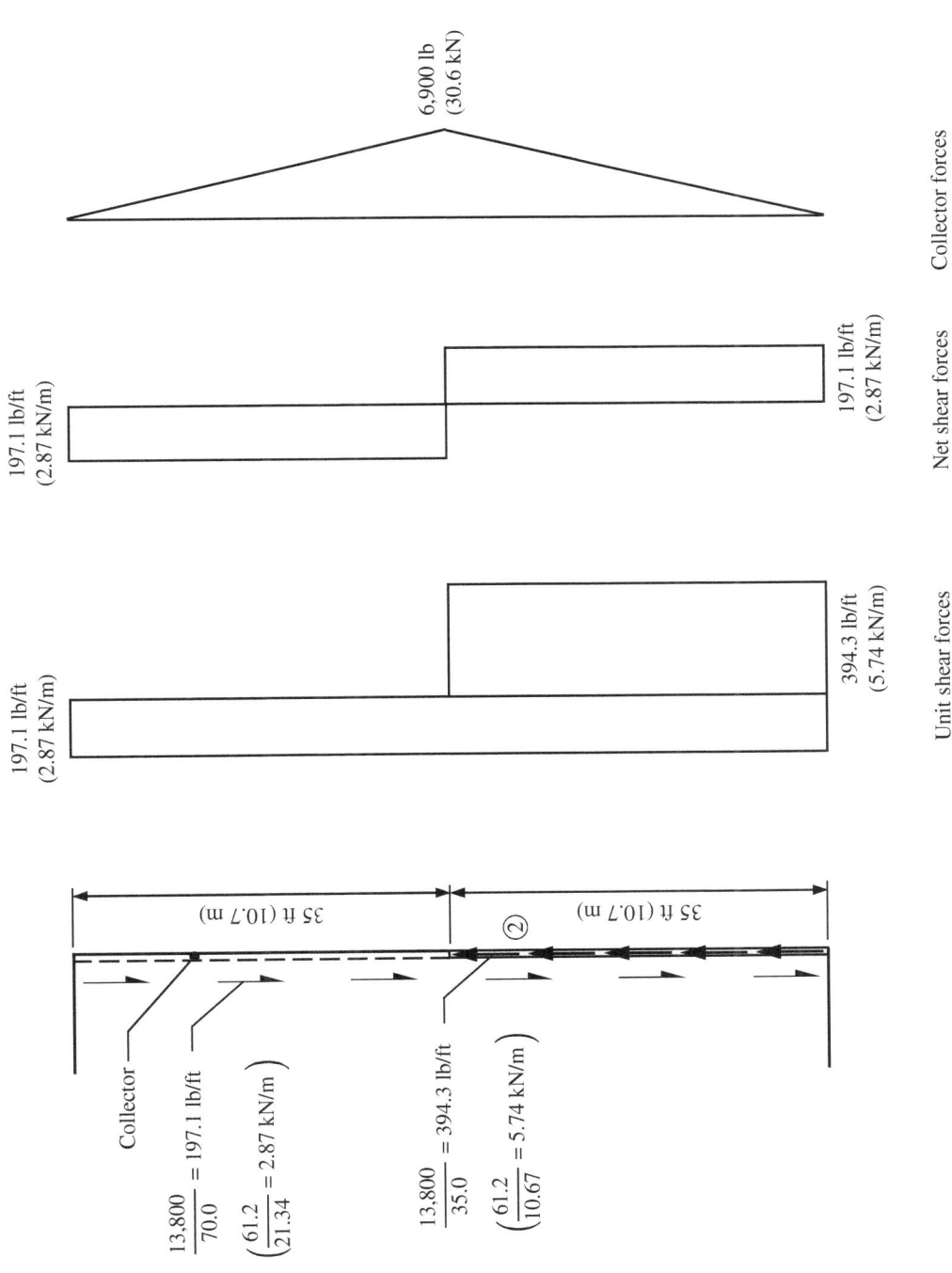

6,900 lb
(30.6 kN)

Collector forces

197.1 lb/ft
(2.87 kN/m)

197.1 lb/ft
(2.87 kN/m)

Net shear forces

197.1 lb/ft
(2.87 kN/m)

394.3 lb/ft
(5.74 kN/m)

Unit shear forces

35 ft (10.7 m)

35 ft (10.7 m)

②

Collector

$\dfrac{13,800}{70.0} = 197.1$ lb/ft

$\left(\dfrac{61.2}{21.34} = 2.87 \text{ kN/m}\right)$

$\dfrac{13,800}{35.0} = 394.3$ lb/ft

$\left(\dfrac{61.2}{10.67} = 5.74 \text{ kN/m}\right)$

Figure 3.41 Unit shear forces, net shear forces, and collector forces in the diaphragm at wall 2 in Example 3.13.

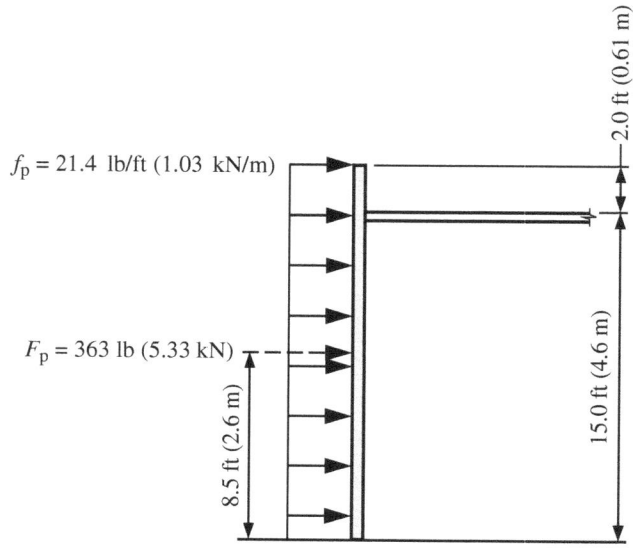

$f_p = 21.4$ lb/ft (1.03 kN/m)

$F_p = 363$ lb (5.33 kN)

8.5 ft (2.6 m)

15.0 ft (4.6 m)

2.0 ft (0.61 m)

Figure 3.42 Out-of-plane seismic forces on the masonry walls in Example 3.14.

The amplification factor for diaphragm flexibility, k_a, is determined by ASCE/SEI Equation (12.11-2) for connections to flexible diaphragms:

$$k_a = 1.0 + \frac{L_f}{100} = 1.0 + \frac{70.0}{100} = 1.7 < 2.0$$

where L_f = span of a flexible diaphragm providing lateral support of the wall = 70.0 ft (21.3 m).

W_p = weight of the wall tributary to the anchor = $86.0 \times [(15.0/2) + 2.0] = 817$ lb/ft

$F_p = 0.4 \times 0.621 \times 1.7 \times 1.0 \times 817 = 345$ lb/ft $> 0.2 \times 1.7 \times 1.0 \times 817 = 278$ lb/ft

For example, if the anchors are spaced 3.0 ft on center, the anchor must be designed for a force equal to $345 \times 3.0 = 1,035$ lb.

In S.I.:

W_p = weight of the wall tributary to the anchor = $4.12 \times [(4.6/2) + 0.61] = 12.0$ kN/m

$F_p = 0.4 \times 0.621 \times 1.7 \times 1.0 \times 12.0 = 5.1$ kN/m $> 0.2 \times 1.7 \times 1.0 \times 12.0 = 4.1$ kN/m

For example, if the anchors are spaced 1.0 m on center, the anchor must be designed for a force equal to $5.1 \times 1.0 = 5.1$ kN.

3.16.16 Example 3.16—Story Drift Limit

For the four-story essential facility in Example 3.5 (see Fig. 3.32), check the story drift limits in the north-south direction. The story displacements obtained from the elastic analysis, δ_{xe}, are given in Table 3.25.

Solution

Step 1—Determine the story displacements and drifts

The deflections determined by ASCE/SEI Equation (12.8-15) must be used to check drift limits:

$$\delta_x = \frac{C_d \delta_{xe}}{I_e}$$

The deflection amplification factor, C_d, for special reinforced concrete moment frames is equal to 5.5 (ASCE/SEI Table 12.2-1).

The importance factor, I_e, is equal to 1.5 for Risk Category IV buildings (ASCE/SEI Table 1.5-2).

The story displacements obtained from the elastic analysis, δ_{xe}, and the deflections δ_x are given in Table 3.37 (see Table 3.25).

The interstory drifts, Δ, are also given in Table 3.37 and are determined by subtracting the displacement at the bottom of the story from the displacement at the top of the story:

$$\Delta = \delta_x - \delta_{x-1}$$

Step 2—Determine the allowable story drifts

Allowable story drifts, Δ_a, are determined from ASCE/SEI Table 12.12-1.

For "all other structures," $\Delta_a = 0.010 h_{sx}$ for buildings assigned to Risk Category IV where h_{sx} is the story height below level x. However, because the SFRS consists solely of moment frames in a building assigned to SDC D, $\Delta_a = 0.010 h_{sx}/\rho$ for any story where ρ is the redundancy factor determined in accordance with ASCE/SEI 12.3.4.2 (ASCE/SEI 12.12.1.1).

It is evident that $\rho = 1.0$ because the SFRS consists of more than two bays of seismic force–resisting perimeter framing on each side of the building in each orthogonal direction at each story (ASCE/SEI 12.3.4.2b).

Therefore, for the first story, $\Delta_a = 0.010 h_{sx}/\rho = 0.010 \times 12.0 \times 12.0/1.0 = 1.44$ in.
For the other stories, $\Delta_a = 0.010 \times 11.0 \times 12.0/1.0 = 1.32$ in.

In S.I.:

First story: $\Delta_a = 0.010 \times 3.7 \times 1,000/1.0 = 37.0$ mm

Other stories: $\Delta_a = 0.010 \times 3.4 \times 1,000/1.0 = 34.0$ mm

Story	δ_{xe}, in. (mm)	δ_x, in. (mm)	Δ, in. (mm)
4	1.52 (38.6)	5.57 (141.5)	0.77 (19.4)
3	1.31 (33.3)	4.80 (122.1)	1.28 (32.6)
2	0.96 (24.4)	3.52 (89.5)	1.65 (41.8)
1	0.51 (13.0)	1.87 (47.7)	1.87 (47.7)

TABLE **3.37** Story Displacements and Drifts due to Seismic Forces in the North-South Direction for the Essential Facility in Example 3.16

It is evident that allowable drift limits are not satisfied in the first and second stories of this building. Thus, the drifts must be decreased, which means the structure must be made laterally stiffer.

3.16.17 Example 3.17—Simplified Alternative Structural Design Criteria for a Simple Bearing Wall Building

Determine the seismic base shear and forces for the two-story commercial building in Fig. 3.43 using the simplified alternative structural design criteria in ASCE/SEI 12.14 given the design data in Table 3.38. Assume the diaphragms are flexible at all levels. The story heights are equal to 12 ft (3.7 m).

Solution

The flowchart in Fig. 3.26 is used to determine the seismic base shear and forces.

Step 1—Check if the 12 limitations in ASCE/SEI 12.14.1.1 are satisfied

A check of the limitations is given in Table 3.39.

Because all 12 limitations of ASCE/SEI 12.14.1.1 are satisfied, the simplified design procedure may be used.

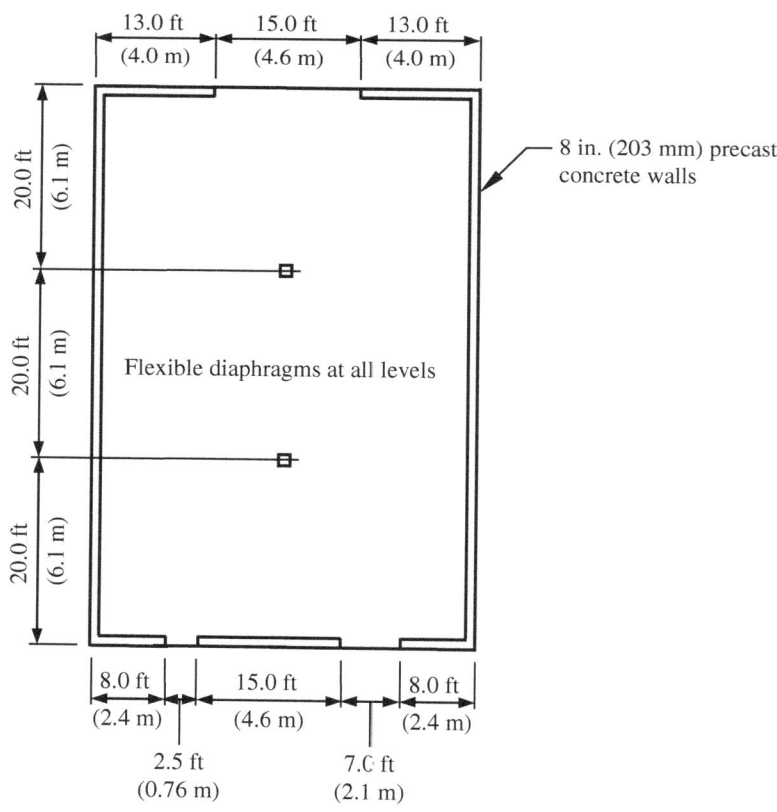

Figure 3.43 Typical plan of the two-story commercial building in Example 3.17.

Location	Latitude: 45.492°, Longitude: −122.635°
Soil classification	Site class C
Occupancy	Commercial occupancy where less than 300 people congregate in one area
SFRS	Bearing wall system with intermediate precast shear walls (system A5 in ASCE/SEI Table 12.14-1)
Weight of roof framing and superimposed dead load	20 lb/ft² (0.96 kN/m²)
Weight of floor framing and superimposed dead load	30 lb/ft² (1.44 kN/m²)
Weight of 8 in. (203 mm) precast shear walls	100 lb/ft² (4.79 kN/m²)

TABLE 3.38 Design Data for the Two-Story Commercial Building in Example 3.17

Step 2—Determine S_S in accordance with ASCE/SEI 11.4.4   ASCE/SEI 12.14.8.1

The mapped spectral response acceleration at short periods, S_S, can be determined from Refs. 5, 6, or 7:

$$S_S = 0.889 < 1.5$$

Step 3—Determine S_{DS}

$$S_{DS} = \frac{2F_a S_S}{3} = \frac{2 \times 1.4 \times 0.889}{3} = 0.830$$

where $F_a = 1.4$ for soil sites.

Note that $F_a = 1.2$ based on the requirements in ASCE/SEI 11.4.4; $F_a = 1.4$ is used in this example.

Step 4—Determine the SDC   Figure 2.6

For Risk Category II and $S_{DS} = 0.830 > 0.50$, the SDC is D.

ASCE/SEI Table 11.6-1

Step 5—Determine the response modification coefficient, R   ASCE/SEI Table 12.14-1

For a bearing wall system with intermediate precast shear walls, $R = 4$.

The height limit for this SFRS is 40.0 ft (12.2 m) for buildings assigned to SDC D, which is greater than the building height of 24.0 ft (7.3 m).

Step 6—Determine the effective seismic weight, W   ASCE/SEI 12.14.8.1

Weight of roof framing and superimposed dead load = $20.0 \times (41.0 \times 60.0)/1,000 = 49$ kips

Weight of floor framing and superimposed dead load = $30.0 \times (41.0 \times 60.0)/1,000 = 74$ kips

Weight of precast walls tributary to the roof and floor diaphragms = $100 \times (12.0 + 6.0) \times [(2 \times 60.0) + 26.0 + 31.0]/1,000 = 319$ kips

$$W = 49 + 74 + 319 = 442 \text{ kips}$$

	Limitation	Discussion
1	The structure must qualify for Risk Category I or II in accordance with ASCE/SEI Table 1.5-1.	The building is assigned to Risk Category II given the commercial occupancy where less than 300 people congregate in one area.
2	The site class defined in ASCE/SEI Chapter 20 must not be site class E or F.	The site class for this building is C (see Table 3.38).
3	The structure must not exceed three stories above the grade plane.	The structure is two stories.
4	The SFRS must be either a bearing wall system or a building frame system, as indicated in ASCE/SEI Table 12.14-1.	The building has a bearing wall system consisting of intermediate precast concrete shear walls (see Table 3.38).
5	The structure must have at least two lines of lateral resistance in each of two major axis directions. At least one line of resistance must be provided on each side of the CM in each direction.	Precast shear walls are provided along two lines at the perimeter of the building in both directions with at least one wall on each side of the CM (see Fig. 3.43).
6	The CM in each story must be located not further from the geometric centroid of the diaphragm than 10 percent of the length of the diaphragm parallel to the eccentricity.	The CM and the geometric centroid of the diaphragm are essentially at the same location.
7	For structures with cast-in-place concrete diaphragms, overhangs beyond the outside line of shear walls or braced frames must satisfy ASCE/SEI Equation (12.14-1).	This limitation is not applicable because cast-in-place concrete diaphragms are not used in this building.
8	For buildings with diaphragms that are not flexible, the forces must be apportioned to the vertical elements of the SFRS as if the diaphragms were flexible.	The diaphragms in this building are flexible.
9	Lines of resistance of the SFRS must be oriented at angles of no more than 15 degrees from alignment with the major orthogonal axes of the building.	The shear walls in both directions are parallel to the major axes.
10	The simplified design procedure must be used for each major horizontal axis direction of the building.	The simplified design procedure is used in both directions.
11	System irregularities caused by in-plane or out-of-plane offsets of the lateral force–resisting elements are not permitted.	This building does not have any irregularities.
12	The lateral load resistance of any story must not be less than 80 percent of the story above.	The lateral load resistances of both stories are equal because the thickness, length, and arrangement of the shear walls are the same in both stories.

TABLE 3.39   Check of the Limitations in ASCE/SEI 12.14.1.1 for the Commercial Building in Example 3.17

In S.I.:

Weight of roof framing and superimposed dead load = $0.96 \times (12.5 \times 18.3) =$ 219.6 kN

Weight of floor framing and superimposed dead load = $1.44 \times (12.5 \times 18.3) =$ 329.4 kN

Weight of precast walls tributary to the roof and floor diaphragms = $4.79 \times (3.7 + 1.8) \times [(2 \times 18.3) + 7.9 + 9.5] = 1,422.6$ kN

$$W = 219.6 + 329.4 + 1,422.6 = 1,971.6 \text{ kN}$$

Step 7—Determine the seismic base shear, V   ASCE/SEI Equation (12.14-12)

$$V = \frac{FS_{DS}}{R}W = \frac{1.1 \times 0.830}{4} \times 442 = 101 \text{ kips}$$

where $F = 1.1$ for a two-story building.

In S.I.:

$$V = \frac{1.1 \times 0.830}{4} \times 1,971.6 = 450 \text{ kN}$$

Step 8—Determine the lateral seismic force, F_x   ASCE/SEI Equation (12.14-13)

$$F_x = \frac{w_x}{W}V$$

- Roof level

$$w_R = 49 + \{100 \times 6.0 \times [(2 \times 60.0) + 26.0 + 31.0]/1,000\} = 155 \text{ kips}$$

$$F_R = \frac{155}{442} \times 101 = 35.4 \text{ kips}$$

In S.I.:

$$w_R = 219.6 + \{4.79 \times 1.8 \times [(2 \times 18.3) + 7.9 + 9.5]\} = 685.2 \text{ kN}$$

$$F_R = \frac{685.2}{1,971.6} \times 450 = 156.4 \text{ kN}$$

- Second floor

$$w_2 = 74 + \{100 \times 12.0 \times [(2 \times 60.0) + 26.0 + 31.0]/1,000\} = 287 \text{ kips}$$

$$F_2 = \frac{287}{442} \times 101 = 65.6 \text{ kips}$$

In S.I.:

$$w_2 = 329.4 + \{4.79 \times 3.7 \times [(2 \times 18.3) + 7.9 + 9.5]\} = 1,286.4 \text{ kN}$$

$$F_2 = \frac{1,286.4}{1,971.6} \times 450 = 293.6 \text{ kN}$$

CHAPTER 4

Seismic Design Requirements for Nonstructural Components

4.1 Overview

Minimum design criteria for nonstructural components permanently attached to structures, including their supports and attachments are given in this chapter, which are based on the provisions in ASCE/SEI Chapter 13. Requirements are provided for architectural components, mechanical and electrical components, and anchorage.

Nonstructural components that weigh greater than or equal to 25 percent of the effective seismic weight, W, of the structure must be designed as nonbuilding structures in accordance with ASCE/SEI 15.3.2 (ASCE/SEI 13.1.1).

Nonstructural components must be assigned to the same SDC as the structure that they occupy or to which they are attached (ASCE/SEI 13.1.2).

The component importance factor, I_p, is equal to 1.0 except where the four conditions in ASCE/SEI 13.1.3 apply, in which it is equal to 1.5.

A list of nonstructural components exempt from the requirements of ASCE/SEI Chapter 13 is given in ASCE/SEI 13.1.4. It is assumed these nonstructural components and systems can achieve the required performance goals due to their inherent strength and stability, the lower level of earthquake demand, or both.

4.2 General Design Requirements

Given in ASCE/SEI Table 13.2-1 are the requirements in ASCE/SEI Chapter 13 applicable to the supports and attachments of architectural, mechanical, and electrical components (see Table 4.1).

4.3 Seismic Demands on Nonstructural Components

4.3.1 Seismic Design Force

Horizontal Force

The horizontal seismic design force, F_p, which is applied to the center of gravity of the component, is determined by ASCE/SEI Equation (13.3-1) with the upper and lower limits determined by ASCE/SEI Equations (13.3-2) and (13.3-3), respectively:

$$0.3S_{DS}I_pW_p \leq F_p = \frac{0.4a_pS_{DS}W_p}{\left(\dfrac{R_p}{I_p}\right)}\left(1+\frac{2z}{h}\right) \leq 1.6S_{DS}I_pW_p \tag{4.1}$$

175

Nonstructural Element	Requirements				
	General Design (ASCE/ SEI 13.2)	**Force and Displacement (ASCE/ SEI 13.3)**	**Attachment (ASCE/ SEI 13.4)**	**Architectural Component (ASCE/SEI 13.5)**	**Mechanical and Electrical Component (ASCE/ SEI 13.6)**
Architectural components and supports and attachments for architectural components	X	X	X	X	
Mechanical and electrical components	X	X	X		X
Supports and attachments for mechanical and electrical components	X	X	X		X

TABLE 4.1 Requirements for Supports and Attachments of Architectural, Mechanical, and Electrical Components

The terms in Eq. (4.1) are defined in Table 4.2.

The force F_p must be applied independently in at least two orthogonal horizontal directions in combination with service or operating loads associated with the component (ASCE/SEI 13.3.1.1). However, for vertically cantilevered systems, F_p must be assumed to act in any horizontal direction.

Term	Definition
a_p	Component amplification factor given in ASCE/SEI Table 13.5-1 for architectural components and ASCE/SEI Table 13.6-1 for mechanical and electrical components
S_{DS}	Design spectral response acceleration at short periods determined in accordance with ASCE/SEI 11.4.5
W_p	Component operating weight
R_p	Component response modification factor given in ASCE/SEI Table 13.5-1 for architectural components and ASCE/SEI Table 13.6-1 for mechanical and electrical components
I_p	Component importance factor, which is equal to 1.0 or 1.5 (see ASCE/SEI 13.1.3)
z	Height in the structure of point of attachment of the component with respect to the base where $z = 0$ for items at or below the base and $z/h \leq 1.0$
h	Average roof height of the structure with respect to the base

TABLE 4.2 Terms in the Determination of the Horizontal Seismic Design Force, F_p

The following conditions are also applicable (see ASCE/SEI 13.3.1.1):

- The overstrength factor, Ω_0, in ASCE/SEI Tables 13.5-1 and 13.6-1 is applicable only to anchorage of components to concrete and masonry where required by ASCE/SEI 13.4.2 or the material standards referenced in ASCE/SEI 13.4.2. The seismic load effects including overstrength in ASCE/SEI 12.4.3 are applicable.

- The redundancy factor, ρ, is permitted to taken as 1.0 and the overstrength factors, Ω_0, in ASCE/SEI Table 12.2-1 are not applicable.

Vertical Force

In accordance with ASCE/SEI 13.3.1.2, components must be designed for a concurrent vertical force equal to $\pm 0.2 S_{DS} W_p$. This force need not be considered for lay-in access floor panels and lay-in ceiling panels.

Nonseismic Loads

In cases where nonseismic loads on nonstructural components are greater than F_p, the component must be designed for the effects of the nonseismic loads; however, the detailing requirements and limitations in ASCE/SEI Chapter 13 must be satisfied.

Dynamic Analysis

In lieu of F_p determined in accordance with ASCE/SEI 13.3.1.1, ASCE/SEI Equation (13.3-4) can be used to calculate F_p based on accelerations determined by the dynamic analysis methods given in ASCE/SEI 13.3.1.4:

$$0.3 S_{DS} I_p W_p \le F_p = \frac{a_i a_p S_{DS} W_p A_x}{\left(\dfrac{R_p}{I_p}\right)} \le 1.6 S_{DS} I_p W_p \tag{4.2}$$

In this equation, a_i is the maximum acceleration at level i obtained from a modal analysis of the structure and A_x is the torsional amplification factor determined by ASCE/SEI Equation (12.8-14) (see Sec. 3.9.5 of this publication). Where a seismic response history analysis with at least seven ground motions is utilized, a_i must be taken as the average of the maximum accelerations; where fewer than seven motions are used, the maximum acceleration value for each floor must be based on the maximum value from the ground motions analyzed.

Floor accelerations at any level are also permitted to be determined using floor response spectra. Requirements for this type of analysis are given in ASCE/SEI 13.3.1.4.1.

4.3.2   Seismic Relative Displacements

The requirements for seismic relative displacements in ASCE/SEI 13.3.2 are applicable in the design of cladding, stairways, windows, piping systems, sprinkler systems, and other components that are connected either to (1) one structure at multiple levels or (2) multiple structures.

Seismic relative displacements, D_{pI}, are determined by ASCE/SEI Equation (13.3-6):

$$D_{pI} = D_p I_e \tag{4.3}$$

In this equation, D_p is the displacement determined in accordance with ASCE/ SEI 13.3.2.1 for displacements within structures and ASCE/SEI 13.3.2.2 for displacements

Displacements within structures (ASCE/SEI 13.3.2.1)	• $D_p = \delta_{xA} - \delta_{ya} \le (h_x - h_y)\Delta_{aA}/h_{sx}$ δ_{xA} = deflection at building level x of structure A determined in accordance with ASCE/SEI Equation (12.8-15) δ_{yA} = deflection at building level y of structure A determined in accordance with ASCE/SEI Equation (12.8-15) h_x = height of level x to which the upper connection point is attached h_y = height of level y to which the lower connection point is attached Δ_{aA} = allowable story drift for structure A determined in accordance with ASCE/SEI Table 12.12-1 h_{sx} = story height used in the definition of the allowable drift, Δ_a, in ASCE/SEI Table 12.12-1 • D_p is permitted to be determined using the linear dynamic procedures in ASCE/SEI 12.9
Displacements between structures (ASCE/SEI 13.3.2.2)	$D_p = \lvert \delta_{xA} \rvert + \lvert \delta_{yB} \rvert \le \dfrac{h_x \Delta_{aA}}{h_{sx}} + \dfrac{h_y \Delta_{aB}}{h_{sx}}$ where δ_{xA} = deflection at building level x of structure A determined in accordance with ASCE/SEI Equation (12.8-15) δ_{yB} = deflection at building level y of structure B determined in accordance with ASCE/SEI Equation (12.8-15) h_x = height of level x to which the upper connection point is attached h_y = height of level y to which the lower connection point is attached Δ_{aA} = allowable story drift for structure A determined in accordance with ASCE/SEI Table 12.12-1 Δ_{aB} = allowable story drift for structure B determined in accordance with ASCE/SEI Table 12.12-1 h_{sx} = story height used in the definition of the allowable drift, Δ_a, in ASCE/SEI Table 12.12-1

TABLE 4.3 Seismic Relative Displacements

between structures (see Table 4.3). Also, I_e is the seismic importance factor for the structure determined in accordance with ASCE/SEI 11.5.1.

An example of seismic relative displacements within a structure is given in Fig. 4.1 for a glazing system. The piping system connecting the two buildings in Fig. 4.2 is an example of seismic relative displacements between two structures.

The effects of seismic relative displacements must be considered in combination with displacements due to other loads.

4.3.3 Component Period

The fundamental period, T_p, of a nonstructural component, including its supports and attachments to the structure, is permitted to be determined by ASCE/SEI Equation (13.3-11):

$$T_p = 2\pi \sqrt{\frac{W_p}{K_p g}} \qquad (4.4)$$

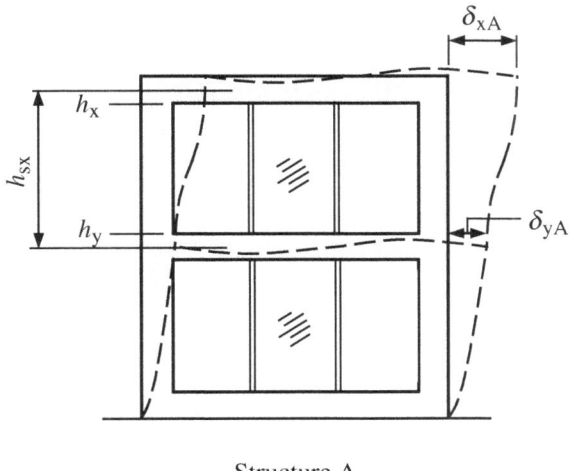

Structure A

FIGURE 4.1   Seismic relative displacements within a structure.

In this equation, K_p is the combined stiffness of the component, supports, and attachments determined in terms of load per deflection at the center of gravity of the component and g is the acceleration due to gravity.

In lieu of ASCE/SEI Equation (13.3-11), T_p is permitted to be determined from experimental test data or by a properly substantiated analysis.

4.4   Nonstructural Component Anchorage

Components and their supports must be attached or anchored to the structure in accordance with the requirements in ASCE/SEI 13.4. Forces in the attachments must be determined using the component forces and displacements given in ASCE/SEI 13.3.1 and 13.3.2, respectively, with the exception that R_p must be taken less than or equal to 6. Anchors in concrete or masonry elements must satisfy the provisions of ASCE/SEI 13.4.2.

4.5   Architectural Components

Design and detailing requirements for architectural components are given in ASCE/SEI 13.5. Architectural components and attachments must be designed for the seismic component forces in ASCE/SEI 13.3.1 (ASCE/SEI 13.5.2). Requirements are given for the following:

- Exterior nonstructural wall elements and connections (ASCE/SEI 13.5.3)
- Glass (ASCE/SEI 13.5.4)
- Out-of-plane bending (ASCE/SEI 13.5.5)
- Suspended ceilings (ASCE/SEI 13.5.6)
- Access floors (ASCE/SEI 13.5.7)

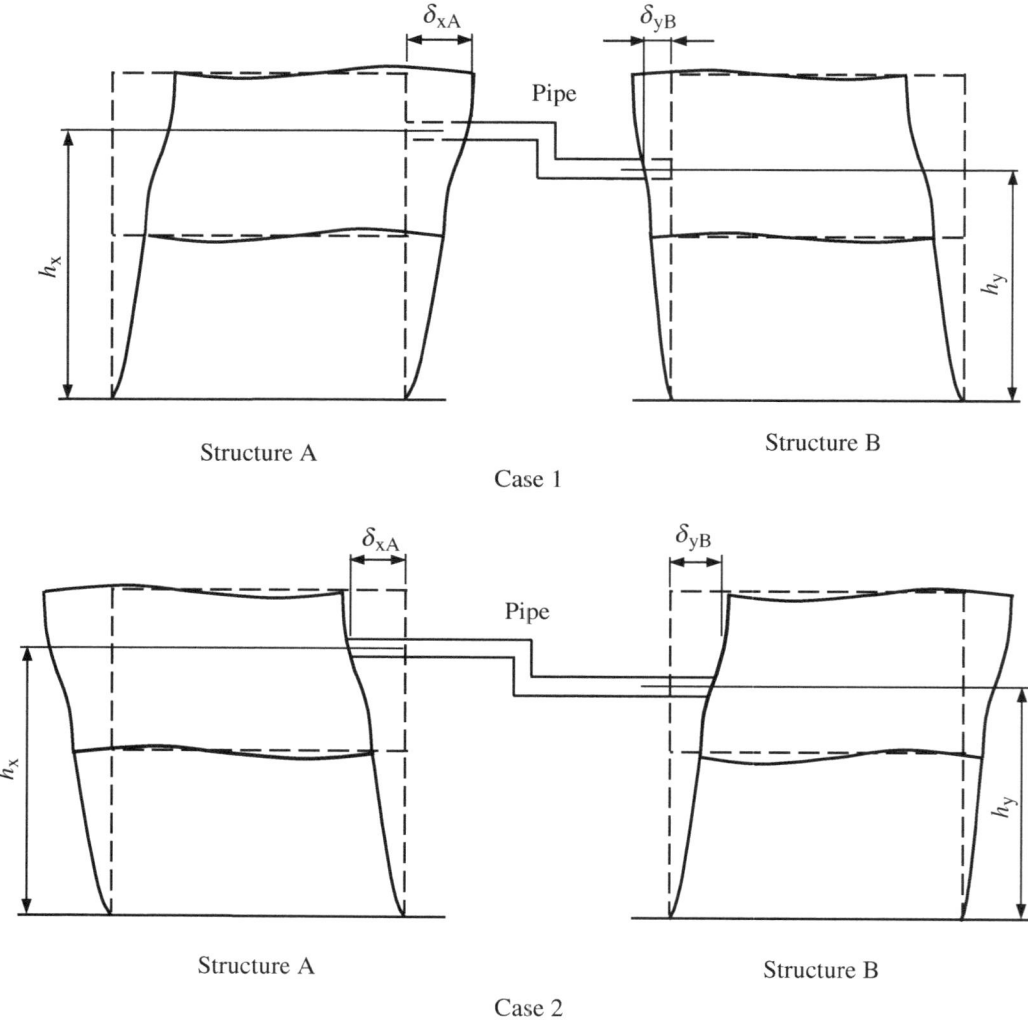

FIGURE 4.2 Seismic relative displacements between structures.

- Partitions (ASCE/SEI 13.5.8)
- Glass in glazed curtain walls, glazed storefronts, and glazed partitions (ASCE/SEI 13.5.9)
- Egress stairs and ramps (ASCE/SEI 13.5.10)

4.6 Mechanical and Electrical Components

The requirements of ASCE/SEI 13.6 must be satisfied for mechanical and electrical components and their supports.

Requirements are provided for the following:

- Mechanical components (ASCE/SEI 13.6.2)
- Electrical components (ASCE/SEI 13.6.3)
- Component supports (ASCE/SEI 13.6.4)
- Distribution systems
  - Conduit, cable trays, and raceways (ASCE/SEI 13.6.5)
  - Duct systems (ASCE/SEI 13.6.6)
  - Piping and tube systems (ASCE/SEI 13.6.7)
  - Trapezes with a combination of systems (ASCE/SEI 13.6.8)
- Utility and service lines (ASCE/SEI 13.6.9)
- Boilers and pressure vessels (ASCE/SEI 13.6.10)
- Elevators and escalators (ASCE/SEI 13.6.11)
- Rooftop solar panels (ASCE/SEI 13.6.12)
- Other mechanical and electrical components (ASCE/SEI 13.6.13)

The flowchart in Fig. 4.3 can be used to determine design seismic forces on nonstructural components.

4.7 Examples

The following examples illustrate the determination of seismic forces and their effects on nonstructural components based on the requirements given in this chapter.

4.7.1 Example 4.1—Architectural Component—Parapet

Determine the design seismic force on the 2-ft (0.61-m) parapet of the one-story commercial building in Example 3.13.

Solution

The flowchart in Fig. 4.3 is used to determine the seismic force on the parapet.

Step 1—Determine S_{DS}, S_{D1}, and the SDC

From Example 3.13:

- $S_{DS} = 0.621$
- $S_{D1} = 0.270$
- SDC is D

Step 2—Determine a_p   ASCE/SEI Table 13.5-1

The parapet in this example is a cantilever element that is not braced to the structural frame below its center of mass; therefore, $a_p = 2.5$.

Step 3—Determine R_p   ASCE/SEI Table 13.5-1

$$R_p = 2.5$$

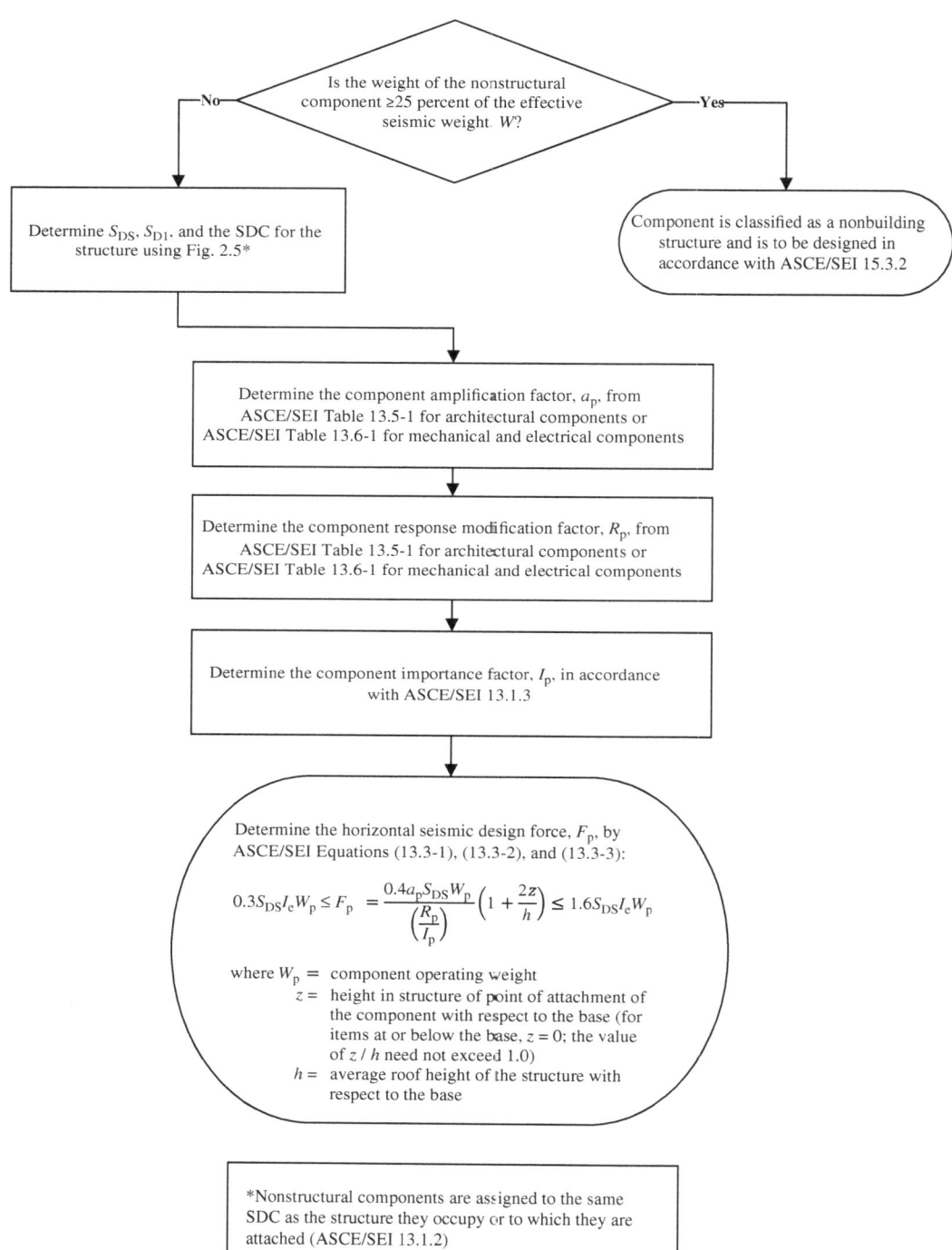

FIGURE 4.3   Flowchart to determine design seismic forces on nonstructural components.

Step 4—Determine I_p   ASCE/SEI 13.1.3

Because the parapet does not meet any of the criteria that require $I_p = 1.5$, $I_p = 1.0$.

Step 5—Determine F_p   ASCE/SEI 13.3.1.1

$$F_p = \frac{0.4a_p S_{DS} W_p}{\left(\dfrac{R_p}{I_p}\right)}\left(1+\frac{2z}{h}\right)$$

Weight of the parapet:

8-in. (203-mm) concrete masonry = 86 lb/ft^2 (4.12 kN/m^2)

$$W_p = 86 \times 2.0 = 172 \text{ lb per foot width of parapet}$$

In S.I.:

$$W_p = 4.12 \times 0.61 = 2.5 \text{ kN per meter width of parapet}$$

Because the parapet is attached to the top of the structure, $z/h = 1.0$.

Therefore,

$$F_p = \frac{0.4 \times 2.5 \times 0.621 \times 172}{\left(\dfrac{2.5}{1.0}\right)} \times (1+2) = 128 \text{ lb per foot width of parapet}$$

Minimum $F_p = 0.3S_{DS}I_p W_p = 0.3 \times 0.621 \times 1.0 \times 172 = 32$ lb per foot width of parapet

Maximum $F_p = 1.6S_{DS}I_p W_p = 1.6 \times 0.621 \times 1.0 \times 172 = 171$ lb per foot width of parapet

In S.I.:

$$F_p = \frac{0.4 \times 2.5 \times 0.621 \times 2.5}{\left(\dfrac{2.5}{1.0}\right)} \times (1+2) = 1.9 \text{ kN per meter width of parapet}$$

Minimum $F_p = 0.3S_{DS}I_p W_p = 0.3 \times 0.621 \times 1.0 \times 2.5 = 0.47$ kN per meter width of parapet

Maximum $F_p = 1.6S_{DS}I_p W_p = 1.6 \times 0.621 \times 1.0 \times 2.5 = 2.5$ kN per meter width of parapet

This out-of-plane force can be expressed as a uniformly distributed load over the height of the parapet (see Fig. 4.4):

$$f_p = 128/2.0 = 64 \text{ lb/ft per foot width of parapet}$$

In S.I.:

$$f_p = 1.9/0.61 = 3.1 \text{ kN/m per meter width of parapet}$$

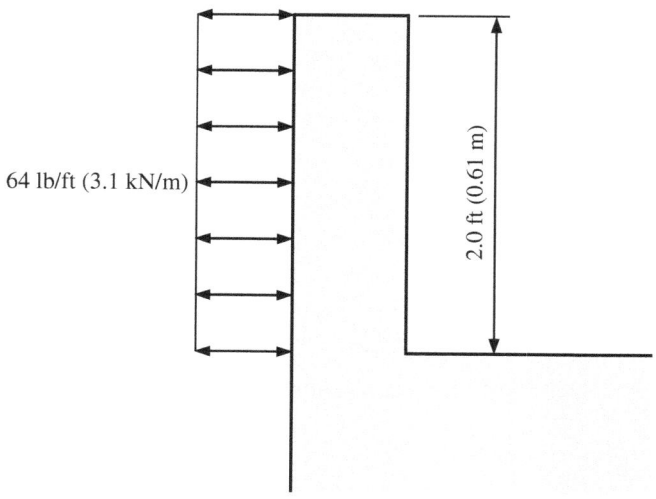

64 lb/ft (3.1 kN/m)

2.0 ft (0.61 m)

FIGURE 4.4   Seismic force on the parapet in Example 4.1.

4.7.2   Example 4.2—Mechanical Component—Air Handling Unit

A 2,000-lb (8.9-kN) air handling unit (AHU) is mounted directly on the roof of the one-story commercial building in Example 3.13. Determine the design seismic force on the AHU.

Solution

The flowchart in Fig. 4.3 is used to determine the seismic force on the AHU.

> *Step 1—Determine S_{DS}, S_{D1}, and the SDC*
> From Example 3.13:
>
> - $S_{DS} = 0.621$
> - $S_{D1} = 0.270$
> - SDC is D

> *Step 2—Determine a_p*   ASCE/SEI Table 13.6-1
> For an AHU, $a_p = 2.5$.

> *Step 3—Determine R_p*   ASCE/SEI Table 13.6-1
> For an AHU, $R_p = 6.0$.

> *Step 4—Determine I_p*   ASCE/SEI 13.1.3
> Because the AHU does not meet any of the criteria that require $I_p = 1.5$, $I_p = 1.0$.

Step 5—Determine F_p   ASCE/SEI 13.3.1.1

$$F_p = \frac{0.4 a_p S_{DS} W_p}{\left(\dfrac{R_p}{I_p}\right)}\left(1 + \frac{2z}{h}\right)$$

Weight of the AHU: $W_p = 2,000$ lb (8.9 kN)

Because the AHU is attached to the top of the structure, $z/h = 1.0$.

Therefore,

$$F_p = \frac{0.4 \times 2.5 \times 0.621 \times 2,000}{\left(\dfrac{6.0}{1.0}\right)}\times(1+2) = 621 \text{ lb}$$

Minimum $F_p = 0.3 S_{DS} I_p W_p = 0.3 \times 0.621 \times 1.0 \times 2,000 = 373$ lb

Maximum $F_p = 1.6 S_{DS} I_p W_p = 1.6 \times 0.621 \times 1.0 \times 2,000 = 1,987$ lb

In S.I.:

$$F_p = \frac{0.4 \times 2.5 \times 0.621 \times 8.9}{\left(\dfrac{6.0}{1.0}\right)}\times(1+2) = 2.8 \text{ kN}$$

Minimum $F_p = 0.3 S_{DS} I_p W_p = 0.3 \times 0.621 \times 1.0 \times 8.9 = 1.7$ kN

Maximum $F_p = 1.6 S_{DS} I_p W_p = 1.6 \times 0.621 \times 1.0 \times 8.9 = 8.8$ kN

The connections between the support of the AHU and the roof deck must be designed for F_p. According to footnote b in ASCE/SEI Table 13.6-1, the design force must be taken as $2F_p$ if the nominal clearance (air gap) between the equipment support frame and the restraint is greater than 0.25 in. (6 mm). Also, according to footnote c in ASCE/SEI Table 13.6-1, if the equipment is to be anchored to concrete or masonry, the force that must be transferred from the equipment to the concrete or masonry must be equal to $\Omega_0 F_p$ where the overstrength factor, Ω_0, is given in ASCE/SEI Table 13.6-1, which is equal to 2 for an AHU.

4.7.3 Example 4.3—Architectural Component—Exterior Nonstructural Wall

Assume the four-story emergency treatment facility in Example 3.5 has cladding on the exterior of the building that consists of 8-in. (203-mm) thick precast concrete wall panels that are 20-ft (6.1 m) wide and that weigh 100 lb/ft² (4.8 kN/m²). Determine the design seismic force on the precast concrete wall panels and their connections to the building in the first and fourth stories assuming the connections to the structure are made at the floor levels at four points (that is, the wall panels span vertically between floors with two connections per floor). The first and fourth story heights are 12.0 ft (3.7 m) and 11.0 ft (3.4 m), respectively.

Solution

The flowchart in Fig. 4.3 is used to determine the seismic force on the precast concrete wall panels and their connections.

Step 1—Determine S_{DS}, S_{D1}, and the SDC

From Example 3.5:

- $S_{DS} = 0.205$
- $S_{D1} = 0.169$
- SDC is D

Step 2—Determine a_p   ASCE/SEI Table 13.5-1

A summary of the component amplification factors, a_p, are given in Table 4.4 for exterior nonstructural wall elements and connections.

Step 3—Determine R_p   ASCE/SEI Table 13.5-1

A summary of the component response modification factors, R_p, are given in Table 4.5 for exterior nonstructural wall elements and connections.

Step 4—Determine I_p   ASCE/SEI 13.1.3

Assuming that failure of the precast concrete wall panels could impair the continued operation of this risk category IV structure, $I_p = 1.5$.

Step 5—Determine F_p   ASCE/SEI 13.3.1.1

$$F_p = \frac{0.4a_p S_{DS} W_p}{\left(\frac{R_p}{I_p}\right)}\left(1 + \frac{2z}{h}\right)$$

Architectural Component	a_p
Wall element	1.00
Body of wall panel connections	1.00
Fasteners of the connecting system	1.25

TABLE 4.4 Component Amplification Factors, a_p, for the Essential Facility in Example 4.3

Architectural Component	R_p
Wall element	2.5
Body of wall panel connections	2.5
Fasteners of the connecting system	1.0

TABLE 4.5 Component Response Modification Factors, R_p, for the Essential Facility in Example 4.3

- Seismic forces on the wall panels and the body of wall panel connections

Weight of the wall panel: $W_p = 100$ lb/ft² (4.8 kN/m²)

For wall panels spanning between two floors, the seismic force on the panels can be taken as the average of the seismic forces at the attachment locations.

For the first story:

$$z_{upper} = 12.0 \text{ ft}$$

$$z_{lower} = 0 \text{ ft}$$

$$F_{p,upper} = \frac{0.4 \times 1.0 \times 0.205 \times 100 \times 20.0 \times 12.0}{\left(\frac{2.5}{1.5}\right)} \times \left(1 + \frac{2 \times 12.0}{45.0}\right) = 1,811 \text{ lb}$$

$$F_{p,lower} = \frac{0.4 \times 1.0 \times 0.205 \times 100 \times 20.0 \times 12.0}{\left(\frac{2.5}{1.5}\right)} \times (1 + 0) = 1,181 \text{ lb}$$

Minimum $F_p = 0.3 S_{DS} I_p W_p = 0.3 \times 0.205 \times 1.5 \times 100 \times 20.0 \times 12.0 = 2,214 \text{ lb}$

Maximum $F_p = 1.6 S_{DS} I_p W_p = 1.6 \times 0.205 \times 1.5 \times 100 \times 20.0 \times 12.0 = 11,808 \text{ lb}$

Minimum F_p is greater than the seismic forces determined at both the top and bottom of the panel; therefore,

$$F_{p,1} = \frac{2,214 + 2,214}{2} = 2,214 \text{ lb}$$

In S.I.:

$$z_{upper} = 3.7 \text{ m}$$

$$z_{lower} = 0 \text{ m}$$

$$F_{p,upper} = \frac{0.4 \times 1.0 \times 0.205 \times 4.8 \times 6.1 \times 3.7}{\left(\frac{2.5}{1.5}\right)} \times \left(1 + \frac{2 \times 3.7}{13.7}\right) = 8.2 \text{ kN}$$

$$F_{p,lower} = \frac{0.4 \times 1.0 \times 0.205 \times 4.8 \times 6.1 \times 3.7}{\left(\frac{2.5}{1.5}\right)} \times (1 + 0) = 5.3 \text{ kN}$$

Minimum $F_p = 0.3 S_{DS} I_p W_p = 0.3 \times 0.205 \times 1.5 \times 4.8 \times 6.1 \times 3.7 = 10.0 \text{ kN}$

Maximum $F_p = 1.6 S_{DS} I_p W_p = 1.6 \times 0.205 \times 1.5 \times 4.8 \times 6.1 \times 3.7 = 53.3 \text{ kN}$

Minimum F_p is greater than the seismic forces determined at both the top and bottom of the panel; therefore,

$$F_{p,1} = \frac{10.0 + 10.0}{2} = 10.0 \text{ kN}$$

For the fourth story:

$$z_{upper} = 45.0 \text{ ft}$$

$$z_{lower} = 34.0 \text{ ft}$$

$$F_{p,upper} = \frac{0.4 \times 1.0 \times 0.205 \times 100 \times 20.0 \times 11.0}{\left(\dfrac{2.5}{1.5}\right)} \times \left(1 + \frac{2 \times 45.0}{45.0}\right) = 3,247 \text{ lb}$$

$$F_{p,lower} = \frac{0.4 \times 1.0 \times 0.205 \times 100 \times 20.0 \times 11.0}{\left(\dfrac{2.5}{1.5}\right)} \times \left(1 + \frac{2 \times 34.0}{45.0}\right) = 2,718 \text{ lb}$$

Minimum $F_p = 0.3 S_{DS} I_p W_p = 0.3 \times 0.205 \times 1.5 \times 100 \times 20.0 \times 11.0 = 2,030 \text{ lb}$

Maximum $F_p = 1.6 S_{DS} I_p W_p = 1.6 \times 0.205 \times 1.5 \times 100 \times 20.0 \times 11.0 = 10,824 \text{ lb}$

Minimum F_p is less than the seismic forces determined at both the top and bottom of the panel; therefore,

$$F_{p,4} = \frac{3,247 + 2,718}{2} = 2,983 \text{ lb} < \text{maximum } F_p = 10,824 \text{ lb}$$

In S.I.:

$$F_{p,upper} = \frac{0.4 \times 1.0 \times 0.205 \times 4.8 \times 6.1 \times 3.4}{\left(\dfrac{2.5}{1.5}\right)} \times \left(1 + \frac{2 \times 13.7}{13.7}\right) = 14.7 \text{ kN}$$

$$F_{p,lower} = \frac{0.4 \times 1.0 \times 0.205 \times 4.8 \times 6.1 \times 3.4}{\left(\dfrac{2.5}{1.5}\right)} \times \left(1 + \frac{2 \times 10.4}{13.7}\right) = 12.3 \text{ kN}$$

Minimum $F_p = 0.3 S_{DS} I_p W_p = 0.3 \times 0.205 \times 1.5 \times 4.8 \times 6.1 \times 3.4 = 9.2 \text{ kN}$

Maximum $F_p = 1.6 S_{DS} I_p W_p = 1.6 \times 0.205 \times 1.5 \times 4.8 \times 6.1 \times 3.4 = 49.0 \text{ kN}$

Minimum F_p is less than the seismic forces determined at both the top and bottom of the panel; therefore,

$$F_{p,4} = \frac{14.7 + 12.3}{2} = 13.5 \text{ kN} < \text{maximum } F_p = 49.0 \text{ kN}$$

- Seismic forces on the fasteners of the connecting system

  The seismic force on the fasteners of the connecting system can be determined based on the distance from the base of the structure to the uppermost fasteners in the story. As noted in the design data, there are four fasteners per panel, so F_p for each fastener is conservatively calculated based on the location of the uppermost fasteners.

- For the first story:

$z = 12.0$ ft

$$F_p = \frac{0.4 \times 1.25 \times 0.205 \times 100 \times 20.0 \times 12.0}{4 \times \left(\dfrac{1.0}{1.5}\right)} \times \left(1 + \frac{2 \times 12.0}{45.0}\right) = 1{,}415 \text{ lb}$$

Minimum $F_p = 0.3 S_{DS} I_p W_p = 0.3 \times 0.205 \times 1.5 \times 100 \times 20.0 \times 12.0 = 2{,}214$ lb

Maximum $F_p = 1.6 S_{DS} I_p W_p = 1.6 \times 0.205 \times 1.5 \times 100 \times 20.0 \times 12.0 = 11{,}808$ lb

Minimum F_p per anchor, which is equal to $2{,}214/4 = 554$ lb, is less than the seismic force determined at the top of the story; therefore, $F_{p,1} = 1{,}415$ lb.
In S.I.:

$z = 3.7$ m

$$F_p = \frac{0.4 \times 1.25 \times 0.205 \times 4.8 \times 6.1 \times 3.7}{4 \times \left(\dfrac{1.0}{1.5}\right)} \times \left(1 + \frac{2 \times 3.7}{13.7}\right) = 6.4 \text{ kN}$$

Minimum $F_p = 0.3 S_{DS} I_p W_p = 0.3 \times 0.205 \times 1.5 \times 4.8 \times 6.1 \times 3.7 = 10.0$ kN

Maximum $F_p = 1.6 S_{DS} I_p W_p = 1.6 \times 0.205 \times 1.5 \times 4.8 \times 6.1 \times 3.7 = 53.3$ kN

Minimum F_p per anchor, which is equal to $10.0/4 = 2.5$ kN, is less than the seismic force determined at the top of the story; therefore, $F_{p,1} = 6.4$ kN.
For the fourth story:

$z = 45.0$ ft

$$F_p = \frac{0.4 \times 1.25 \times 0.205 \times 100 \times 20.0 \times 11.0}{4 \times \left(\dfrac{1.0}{1.5}\right)} \times \left(1 + \frac{2 \times 45.0}{45.0}\right) = 2{,}537 \text{ lb}$$

Minimum $F_p = 0.3 S_{DS} I_p W_p = 0.3 \times 0.205 \times 1.5 \times 100 \times 20.0 \times 11.0 = 2{,}030$ lb

Maximum $F_p = 1.6 S_{DS} I_p W_p = 1.6 \times 0.205 \times 1.5 \times 100 \times 20.0 \times 11.0 = 10{,}824$ lb

Minimum F_p per anchor, which is equal to $2{,}030/4 = 508$ lb, is less than the seismic force determined at the top of the story; therefore, $F_{p,4} = 2{,}537$ lb.
In S.I.:

$z = 13.7$ m

$$F_p = \frac{0.4 \times 1.25 \times 0.205 \times 4.8 \times 6.1 \times 3.4}{4 \times \left(\dfrac{1.0}{1.5}\right)} \times \left(1 + \frac{2 \times 13.7}{13.7}\right) = 11.5 \text{ kN}$$

Minimum $F_p = 0.3 S_{DS} I_p W_p = 0.3 \times 0.205 \times 1.5 \times 4.8 \times 6.1 \times 3.4 = 9.2$ kN

Maximum $F_p = 1.6 S_{DS} I_p W_p = 1.6 \times 0.205 \times 1.5 \times 4.8 \times 6.1 \times 3.4 = 49.0$ kN

Minimum F_p per anchor, which is equal to $9.2/4 = 2.3$ kN, is less than the seismic force determined at the top of the story; therefore, $F_{p,4} = 11.5$ kN.

4.7.4 Example 4.4–Architectural Component–Egress Stairway

Determine the design seismic forces on a precast concrete egress stairway and its connections in the third story of a four-story precast concrete parking structure (see Fig. 4.5). The stairway is not part of the SFRS. The top of the stairway is connected to a precast beam where the connection resists forces in the vertical and horizontal directions. The bottom of the stairway is also connected to a precast beam where the connection resists forces in the vertical direction and in the horizontal direction perpendicular to the stair; the connection permits movement in the horizontal direction parallel to the stair, which means horizontal forces cannot be resisted in this direction. The parking structure, which has typical story heights of 12 ft-3 in. (3.7 m), is assigned to SDC D where $S_{DS} = 0.730$ and $S_{D1} = 0.450$. The weight of the stairway is 9.6 kips (42.7 kN) and the superimposed dead load over the entire stair run is equal to 1.0 kip (4.5 kN).

Solution

The flowchart in Fig. 4.3 is used to determine the seismic force on the precast concrete egress stairway and its connections.

> *Step 1—Determine S_{DS}, S_{D1}, and the SDC*
>
>   From the design data:

- $S_{DS} = 0.730$
- $S_{D1} = 0.450$
- SDC is D

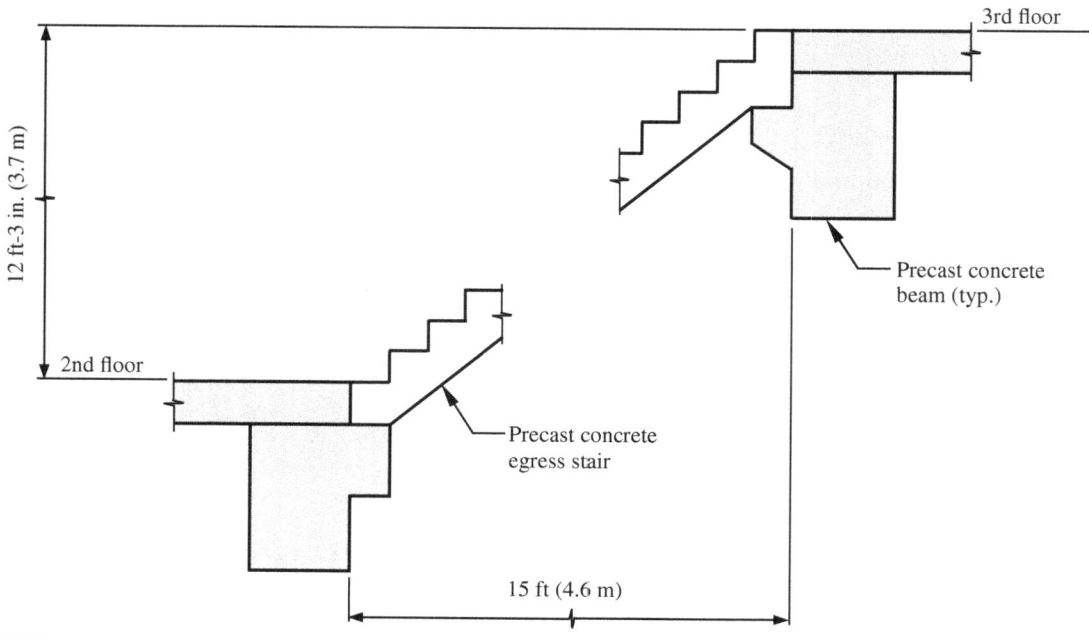

FIGURE 4.5 Precast concrete egress stairway in Example 4.4.

Architectural Component	a_p
Stairway	1.00
Fasteners and attachments	2.50

TABLE 4.6 Component Amplification Factors, a_p, for the Egress Stairway in Example 4.4

Architectural Component	R_p
Stairway	2.50
Fasteners and attachments	2.50

TABLE 4.7 Component Response Modification Factors, R_p, for the Egress Stairway in Example 4.4

Step 2—Determine a_p   ASCE/SEI Table 13.5-1

A summary of the component amplification factors, a_p, are given in Table 4.6.

Step 3—Determine R_p   ASCE/SEI Table 13.5-1

A summary of the component response modification factors, R_p, are given in Table 4.7.

Step 4—Determine I_p   ASCE/SEI 13.1.3

Because this is an egress stairway, $I_p = 1.5$.

Step 5—Determine F_p   ASCE/SEI 13.3.1.1

$$F_p = \frac{0.4 a_p S_{DS} W_p}{\left(\dfrac{R_p}{I_p}\right)}\left(1 + \frac{2z}{h}\right)$$

A summary of the design seismic forces on the egress stairway and the fasteners and attachments is given in Table 4.8 where $h = 49.0$ ft (14.9 m).

Component	W_p, kips (kN)	a_p	R_p	z, ft (m)	F_p, kips (kN)
Perpendicular to the stair					
Stairway	10.6 (47.2)	1.00	2.50	36.75 (11.2)	4.6 (20.7)
Fasteners and attachments	10.6 (47.2)	2.50	2.50	36.75 (11.2)	11.6 (51.8)
Parallel to the stair					
Stairway	10.6 (47.2)	1.00	2.50	30.7 (9.3 m)	4.2 (18.6)
Fasteners and attachments	10.6 (47.2)	2.50	2.50	30.7 (9.3 m)	10.5 (46.5)

TABLE 4.8 Design Seismic Forces on the Egress Stairway and Connections in Example 4.4

$$\text{Minimum } F_p = 0.3S_{DS}I_pW_p = 0.3\times0.730\times1.5\times10.6 = 3.5 \text{ kips}$$

$$\text{Maximum } F_p = 1.6S_{DS}I_pW_p = 1.6\times0.730\times1.5\times10.6 = 18.6 \text{ kips}$$

In S.I.:

$$\text{Minimum } F_p = 0.3S_{DS}I_pW_p = 0.3\times0.730\times1.5\times47.2 = 15.5 \text{ kN}$$

$$\text{Maximum } F_p = 1.6S_{DS}I_pW_p = 1.6\times0.730\times1.5\times47.2 = 82.7 \text{ kips}$$

In the direction perpendicular to the stair, $z = 36.75$ ft (11.2 m) because the only horizontal connection between the stair and the precast beam is at the third elevated floor. In the direction parallel to the stair, horizontal connections are at the second and third elevated floors, so an average z is used, that is, $z = (36.75 + 24.5)/2 = 30.7$ ft (9.3 m).

Calculated values of F_p for fasteners and attachments are multiplied by the overstrength factor, Ω_0, for nonductile anchorage to concrete (see footnote b in ASCE/SEI Table 13.5-1) where $\Omega_0 = 2.5$ for fasteners and attachments of egress stairs (see ASCE/SEI Table 13.5-1).

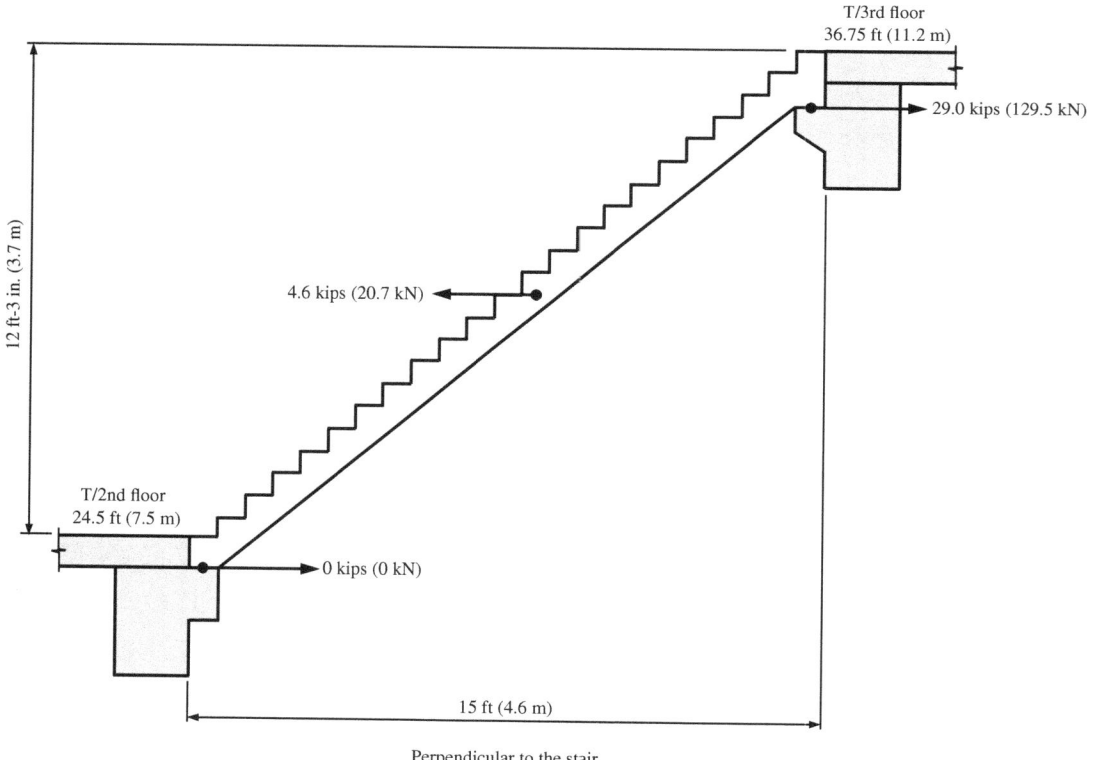

FIGURE 4.6 Design seismic forces on the egress stairway in Example 4.4.

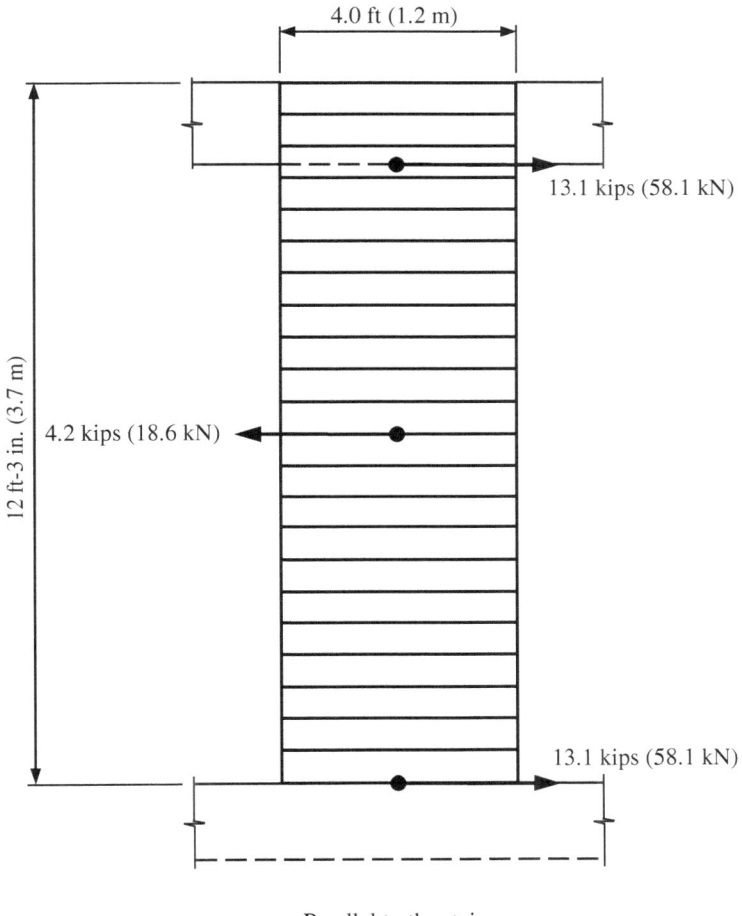

4.0 ft (1.2 m)

13.1 kips (58.1 kN)

12 ft-3 in. (3.7 m)

4.2 kips (18.6 kN)

13.1 kips (58.1 kN)

Parallel to the stair

FIGURE 4.6 *(Continued)*

- Forces perpendicular to the stair:

  Connection force at the top of the stair $= \Omega_0 F_p = 2.5 \times 11.6 = 29.0$ kips

  In S.I.:

  Connection force at the top of the stair $= 2.5 \times 51.8 = 129.5$ kN

- Forces parallel to the stair:

  Connection force at the top and bottom of the stair $= \Omega_0 F_p/2 = 2.5 \times 10.5/2 = 13.1$ kips

  In S.I.:

  Connection force at the top and bottom of the stair $= 2.5 \times 46.5/2 = 58.1$ kN
  The design seismic forces on the egress stair are given in Fig. 4.6.

CHAPTER 5

Seismic Design Requirements for Nonbuilding Structures

5.1 Overview

Nonbuilding structures supported by the ground or by other structures must be designed and detailed to resist the minimum seismic forces in ASCE/SEI Chapter 15.

The selection of a structural analysis procedure for a nonbuilding structure is based on its similarity to buildings. Nonbuilding structures similar to buildings exhibit behavior similar to that of building structures; however, their function and performance are different. According to ASCE/SEI 15.1.3, structural analysis procedures for such buildings are to be selected in accordance with ASCE/SEI 12.6 and Table 12.6-1, which are applicable to building structures. Guidelines and recommendations on the use of these methods are given in ASCE/SEI C15.1.3. The requirements for building structures need to be carefully examined before they are applied to nonbuilding structures.

Nonbuilding structures not similar to buildings exhibit behavior that is markedly different than that of building structures. Most of these types of structures have reference documents that address their unique structural performance and behavior. Such reference documents are permitted to be used to analyze the structure (ASCE/SEI 15.1.3). In addition, the following procedures may be used: equivalent lateral force procedure (ASCE/SEI 12.8), linear dynamic analysis procedures (ASCE/SEI 12.9), and nonlinear response history analysis procedure (ASCE/SEI Chapter 16). Guidelines and recommendations on the proper analysis method to utilize for nonbuilding structures not similar to buildings are given in ASCE/SEI C15.1.3.

Requirements for nonbuilding structures sensitive to vertical ground motion, including tanks, vessels, hanging structures, and nonbuilding structures incorporating horizontal cantilevers, are given in ASCE/SEI 15.1.4.

5.2 Nonbuilding Structures Supported by Other Structures

Requirements are given in ASCE/SEI 15.3 for the nonbuilding structures not similar to buildings identified in ASCE/SEI Table 15.4-2, which are supported by other structures and which are not part of the SFRS. The design method to be used depends on the weight of the nonbuilding structure relative to the weight of the combined nonbuilding and supporting structure:

- Where the weight of the nonbuilding structure is less than 25 percent of the combined effective seismic weights of the nonbuilding structure and the supporting structure, the design seismic forces of the nonbuilding structure are to be determined in accordance with the nonstructural component requirements

197

Fundamental Period, T, of the Nonbuilding Structure	Analysis Requirements
$T < 0.06$ s	• The nonbuilding structure is considered a rigid element with appropriate distribution of its effective seismic weight. • The supporting structure must be designed using the appropriate requirements of ASCE/SEI Chapter 12 or ASCE/SEI 15.5. • The response modification coefficient, R, of the combined system is permitted to be taken as R of the supporting system. • The nonbuilding structure and its attachments must be designed in accordance with ASCE/SEI Chapter 13 for nonstructural components where the component response modification factor, R_p, must be taken equal to R of the nonbuilding structure given in ASCE/SEI Table 15.4-2 and the component amplification factor, a_p, must be taken as 1.0.
$T \geq 0.06$ s	• The nonbuilding structure and supporting structure must be modeled together in a combined model with appropriate stiffness and effective seismic weight distributions. • The combined system must be designed in accordance with the requirements for nonbuilding structures similar to buildings in ASCE/SEI 15.5 with R of the combined system taken as the lesser of R the nonbuilding structure or R of the supporting structure. • The nonbuilding structure and attachments must be designed for the forces determined for the nonbuilding structure in the combined analysis.

TABLE 5.1 Analysis Requirements in Accordance with ASCE/SEI 15.3.2

in ASCE/SEI Chapter 13 (ASCE/SEI 15.3.1; see Chap. 4 of this publication). The appropriate requirements in ASCE/SEI Chapter 12 or ASCE/SEI 15.5 are to be used in the design of the supporting structure.

• Where the weight of the nonbuilding structure is greater than or equal to 25 percent of the combined effective seismic weights of the nonbuilding structure and the supporting structure, an analysis combining the structural characteristics of both methods must be performed, which is based on the fundamental period, T, of the nonbuilding structure (see ASCE/SEI 15.3.2 and Table 5.1).

Architectural, mechanical, and electrical components supported by nonbuilding structures must be designed in accordance with the requirements for nonstructural components in ASCE/SEI Chapter 13 (ASCE/SEI 15.3.3).

5.3 Structural Design Requirements

5.3.1 Design Basis

Basic coefficients and minimum design forces to be used in the determination of design seismic forces for nonbuilding structures are given in ASCE/SEI 15.4. Also provided are height limits and restrictions for nonbuilding structures.

Nonbuilding structures must be designed in accordance with the requirements in ASCE/SEI 15.5 for nonbuilding structures similar to buildings or ASCE/SEI 15.6 for nonbuilding structures not similar to buildings. The seismic lateral forces determined by these requirements must be greater than or equal to those determined by the equivalent lateral force (ELF) procedure in ASCE/SEI 12.8 considering the additions and exceptions in ASCE/SEI 15.4.1, which are summarized in Table 5.2.

5.3.2 Rigid Nonbuilding Structures

As noted previously, rigid nonbuilding structures have a fundamental period, T, less than 0.06 s. The total design lateral seismic base shear, V, in such cases is determined by ASCE/SEI Equation (15.4-5):

$$V = 0.30 S_{DS} W I_e \tag{5.1}$$

where W is the operating weight of the nonbuilding structure.

The base shear is distributed over the height of the structure in accordance with ASCE/SEI 12.8.3.

5.3.3 Loads

The seismic effective weight, W, for nonbuilding structures must include the following where applicable (ASCE/SEI 15.4.3):

- Dead load and other loads defined in ASCE/SEI 12.7.2.
- Weight of normal operating contents for items such as tanks, vessels, bins, hoppers, and the contents of piping.
- Snow and ice loads where these loads constitute 25 percent or more of W or where required by the authority having jurisdiction based on local environmental characteristics.

5.3.4 Fundamental Period

In lieu of determining the fundamental period, T, of the nonbuilding structure using a substantiated dynamic analysis, it is permitted to calculate T by ASCE/SEI Equation (15.4-6):

$$T = 2\pi \sqrt{\frac{\sum w_i \delta_i^2}{g \sum f_i \delta_i}} \tag{5.2}$$

In this equation, f_i represent any set of lateral forces applied over the height of the structure and δ_i are the elastic deflections caused by those forces.

ASCE/SEI Equations (12.8-7) through (12.8-10) are not permitted to be used for determining T (ASCE/SEI 15.4.4).

5.3.5 Drift Limit

For nonbuilding structures, deflection, drifts, and structure separation are calculated using strength design factored load combinations in order to be compatible with the prescribed definition of the seismic load and the definition of the deflection amplification factors in ASCE/SEI Tables 15.4-1 and 15.4-2.

Additions and Exceptions in ASCE/SEI 15.4.1	Requirement
Selection of SFRS	• Nonbuilding structures similar to buildings 1. An SFRS must be selected from those given in ASCE/SEI Table 15.4-1 based on SDC and subject to the limitations given in the table. 2. The seismic coefficients given in ASCE/SEI Table 15.4-1 must be used to determine the seismic base shear, V, element design forces, and design story drifts. 3. Design and detailing requirements must comply with the sections referenced in ASCE/SEI Table 15.4-1. • Nonbuilding structures not similar to buildings 1. An SFRS must be selected from those given in ASCE/SEI Table 15.4-2 based on SDC and subject to the limitations given in the table. 2. The seismic coefficients given in ASCE/SEI Table 15.4-2 must be used to determine the seismic base shear, V, element design forces, and design story drifts. 3. Design and detailing requirements must comply with the sections referenced in ASCE/SEI Table 15.4-2. • Nonbuilding structures not covered in ASCE/SEI Tables 15.4-1 and 15.4-2 1. Applicable strength and other design criteria must be obtained from a reference document applicable to the specific nonbuilding structure. 2. Design and detailing requirements must comply with the reference standard.
Minimum seismic response coefficient, C_s	• For nonbuilding structure types in ASCE/SEI Table 15.4-2, the minimum C_s determined by ASCE/SEI Equation (12.8-5) must be replaced by that determined by ASCE/SEI Equation (15.4-1): $$C_s = 0.044S_{DS}I_e \geq 0.03.$$ • For nonbuilding structures located where $S_1 \geq 0.6$, the minimum C_s determined by ASCE/SEI Equation (12.8-6) must be replaced by that determined by ASCE/SEI Equation (15.4-2): $$C_s = 0.8S_1/(R/I_e).$$ • See the exception in ASCE/SEI 15.4.1(2) for tanks and vessels designed in accordance with the cited reference documents.
Seismic importance factor, I_e	The value of I_e must be the largest value determined by the following: • Applicable reference document in ASCE/SEI Chapter 23. • The largest value from ASCE/SEI Table 1.5-2 based on risk category. • As specified elsewhere in ASCE/SEI Chapter 15.

TABLE 5.2 Structural Design Requirements for Nonbuilding Structures

Additions and Exceptions in ASCE/SEI 15.4.1	Requirement
Vertical distribution of seismic forces	Vertical distribution of lateral seismic forces must be determined by one of the following: • The requirements in ASCE/SEI 12.8.3. • The procedures in ASCE/SEI 12.9.1. • In accordance with the applicable reference document.

Accidental torsion	The accidental torsion requirements of ASCE/SEI 12.8.4.2 need not be accounted for the following structures provided the mass locations for the structure, any contents, and any supported structural or nonstructural elements (including but not limited to piping and stairs) that could contribute to the mass or stiffness of the structure are accounted for and quantified in the analysis:
	• Rigid nonbuilding structures.
	• Nonbuilding structures not similar to buildings with R values less than or equal to 3.5.
	• Nonbuilding structures similar to buildings with R values less than or equal to 3.5 provided one of the following conditions is met:
	  1. The calculated CR at each diaphragm is greater than 5 percent of the plan dimension of the diaphragm in each direction from the calculated CM.
	  2. The structure does not have a horizontal torsional irregularity type 1A or 1B in ASCE/SEI Table 12.3-1 and the structure has at least two lines of lateral resistance in each of two major axis directions. At least one line of lateral resistance must be provided at a distance greater than or equal to 20 percent of the structure's plan dimension from the CM on each side of the CM.
	Structures designed in accordance with the above requirements must be analyzed using a three-dimensional representation of the structure in accordance with ASCE/SEI 12.7.3.

Minimum design seismic force for nonbuilding structures containing liquids, gases, and granular solids supported at the base	The minimum design seismic force must be greater than or equal to that required by the applicable reference document.
Applicability of reference documents	Where a reference document provides a basis for earthquake-resistant design of a particular type of nonbuilding structure covered by ASCE/SEI Chapter 15, the reference document is not permitted to be used unless the following limitations are satisfied: • The seismic ground accelerations and seismic coefficients must be in conformance with the requirements in ASCE/SEI 11.4. • The values for total lateral force and total base overturning moment must not be less than 80 percent of the base shear and overturning moment values, each adjusted for soil-structure interaction obtained from the applicable requirements in ASCE/SEI 7-16.

TABLE 5.2 Structural Design Requirements for Nonbuilding Structures (*Continued*)

Additions and Exceptions in ASCE/SEI 15.4.1	Requirement
Reduction of design seismic base shear	The design seismic base shear, V, is permitted to be reduced in accordance with ASCE/SEI 19.2 to account for the effects of foundation damping from soil-structure interaction. The reduced base shear must not be taken less than $0.7V$.
Load combinations	The effects on the nonbuilding structure caused by gravity loads and seismic forces must be combined in accordance with the factored load combinations in ASCE/SEI 2.3, unless otherwise noted in ASCE/SEI Chapter 15.
Seismic load effects including overstrength	The design seismic forces on nonbuilding structures must be determined in accordance with ASCE/SEI 12.4.3 where specifically required by ASCE/SEI Chapter 15.

TABLE 5.2   Structural Design Requirements for Nonbuilding Structures (*Continued*)

According to ASCE/SEI 15.4.5, the drift limits in ASCE/SEI 12.12.1 need not apply to nonbuilding structures where a rational analysis indicates larger drifts can be exceeded without adversely affecting structural stability or any attached or interconnected components and elements (such as walkways and piping).

P-delta effects must be considered where applicable and must be based on displacements determined by an elastic analysis multiplied by (C_d / I_e) using the appropriate C_d from ASCE/SEI Tables 12.2-1, 15.4-1, and 15.4-2.

5.3.6   Site-Specific Response Spectra

Site-specific analyses must be utilized for specific types of nonbuilding structures where required by a reference document or the authority having jurisdiction (ASCE/SEI 15.4.8). The analysis must be developed in accordance with ASCE/SEI Chapter 21 and must account for local seismicity and geology, expected recurrence intervals, and magnitudes of seismic events from known seismic hazards.

5.3.7   Anchors in Concrete and Masonry

Requirements for anchors in concrete and masonry used for nonbuilding structure anchorage are given in Table 5.3 (ASCE/SEI 15.4.9).

In lieu of designing masonry anchors in accordance with TMS 402, it permitted to use either of the exceptions in ASCE/SEI 15.4.9.2.

Requirements for post-installed anchors and ASTM F1554 anchors are given in ASCE/SEI 15.4.9.3 and 15.4.9.4, respectively.

Material	Requirement
Concrete	Design in accordance with ACI 318 (Ref. 11)
Masonry	Design in accordance with TMS 402 (Ref. 12)

TABLE 5.3   Anchors in Concrete and Masonry

5.3.8 Requirements for Nonbuilding Structure Foundations on Liquefiable Sites

Where nonbuilding structures are located on liquefiable sites, the foundations must comply with the requirements in ASCE/SEI 12.13.9 and 15.4.10.1 (ASCE/SEI 15.4.10).

5.4 Nonbuilding Structures Similar to Buildings

Certain nonbuilding structures exhibit behavior similar to that of buildings. However, their functions and occupancies are different.

ASCE/SEI 15.5 contains requirements for the following structures, which must be satisfied in addition to the design and detailing requirements outlined previously:

- Pipe racks (ASCE/SEI 15.5.2)
- Storage racks (ASCE/SEI 15.5.3)
- Electrical power generating facilities (ASCE/SEI 15.5.4)
- Structural towers for tanks and vessels (ASCE/SEI 15.5.5)
- Piers and wharves (ASCE/SEI 15.5.6)

5.5 Nonbuilding Structures Not Similar to Buildings

Nonbuilding structures not similar to buildings exhibit behavior significantly different from that of building structures. Most of these structures have reference documents that address their structural performance and behavior.

ASCE/SEI 15.6 contains requirements for the following structures, which must be satisfied in addition to the design and detailing requirements outlined previously:

- Earth-retaining structures (ASCE/SEI 15.6.1)
- Chimneys and stacks (ASCE/SEI 15.6.2)
- Amusement structures (ASCE/SEI 15.6.3)
- Special hydraulic structures (ASCE/SEI 15.6.4)
- Secondary containment systems (ASCE/SEI 15.6.5)
- Telecommunication towers (ASCE/SEI 15.6.6)
- Steel tubular support structures for onshore wind turbine generator systems (ASCE/SEI 15.6.7)
- Ground-supported cantilever walls or fences (ASCE/SEI 15.6.8)

5.6 Tanks and Vessels

Comprehensive seismic design requirements are given in ASCE/SEI 15.7 for tanks, vessels, bins, silos, and similar containers storing liquids, gases, and granular solids supported at the base.

5.7 Design Procedure

The flowchart in Fig. 5.1 can be used to determine the design seismic forces on nonbuilding structures similar to buildings and not similar to buildings.

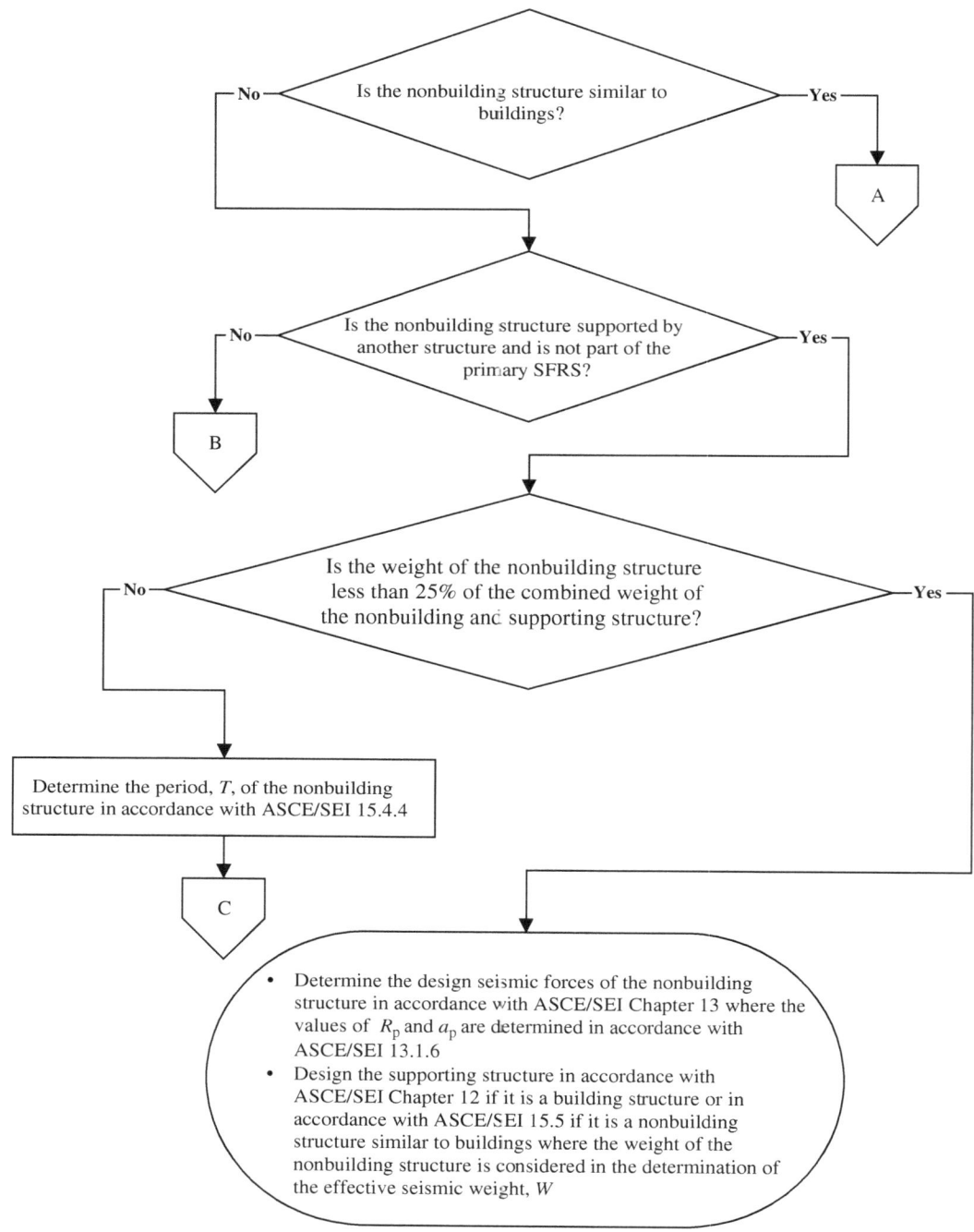

FIGURE 5.1 Flowchart to determine design seismic forces on nonbuilding structures.

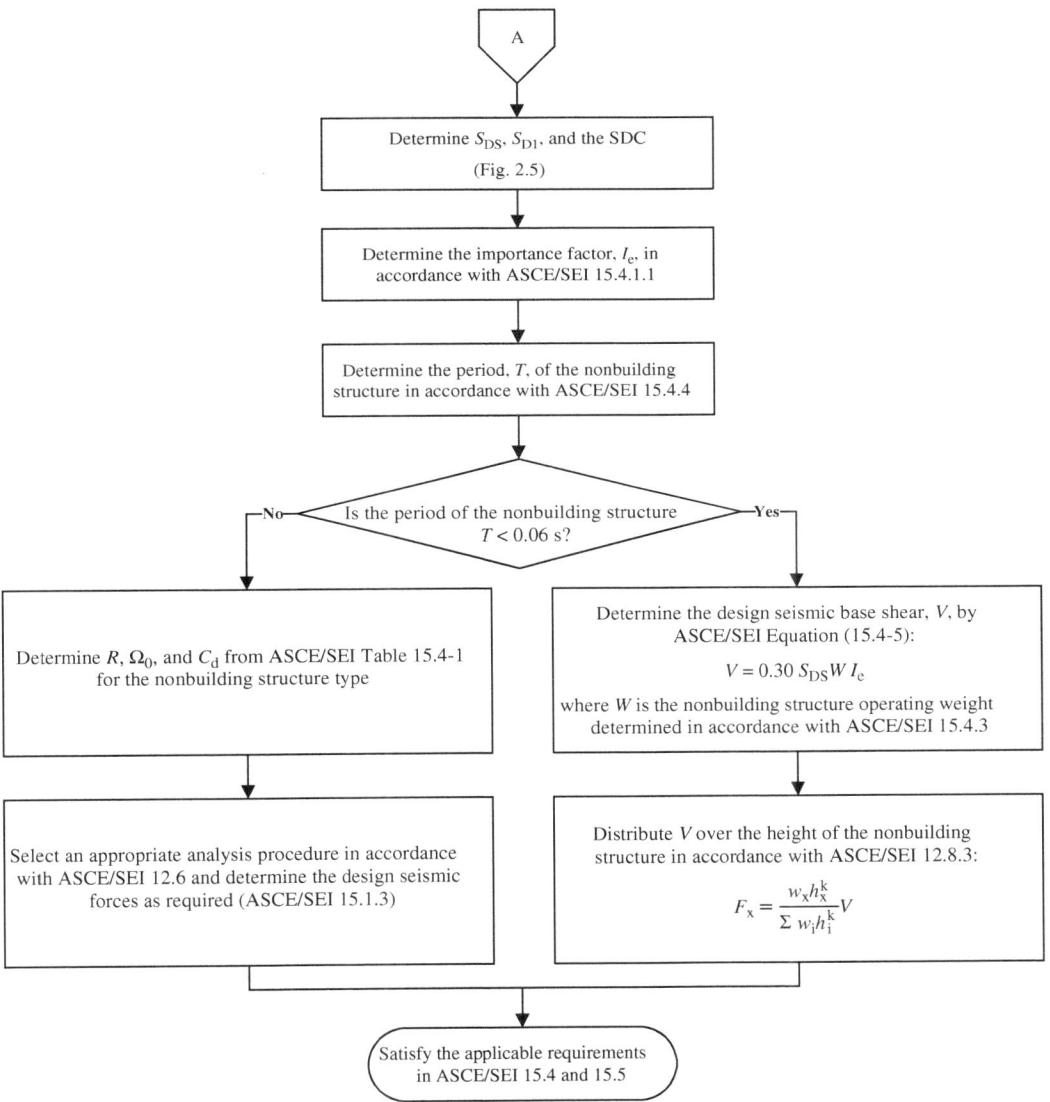

A

Determine S_{DS}, S_{D1}, and the SDC
(Fig. 2.5)

Determine the importance factor, I_e, in
accordance with ASCE/SEI 15.4.1.1

Determine the period, T, of the nonbuilding
structure in accordance with ASCE/SEI 15.4.4

Is the period of the nonbuilding structure
$T < 0.06$ s?

No

Yes

Determine R, Ω_0, and C_d from ASCE/SEI Table 15.4-1
for the nonbuilding structure type

Determine the design seismic base shear, V, by
ASCE/SEI Equation (15.4-5):

$$V = 0.30\, S_{DS} W I_e$$

where W is the nonbuilding structure operating weight
determined in accordance with ASCE/SEI 15.4.3

Select an appropriate analysis procedure in accordance
with ASCE/SEI 12.6 and determine the design seismic
forces as required (ASCE/SEI 15.1.3)

Distribute V over the height of the nonbuilding
structure in accordance with ASCE/SEI 12.8.3:

$$F_x = \frac{w_x h_x^k}{\Sigma\, w_i h_i^k} V$$

Satisfy the applicable requirements
in ASCE/SEI 15.4 and 15.5

FIGURE 5.1 *(Continued)*

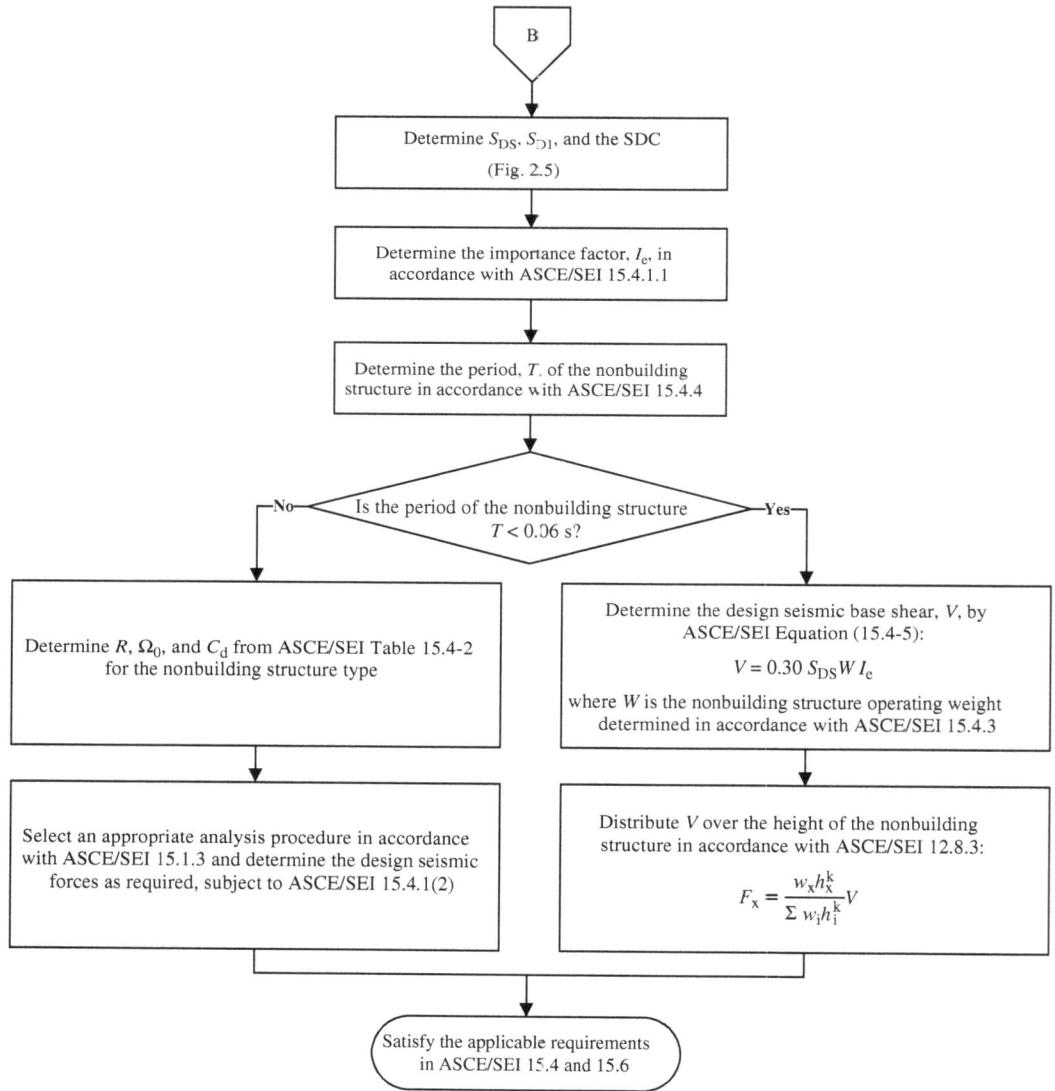

FIGURE 5.1 *(Continued)*

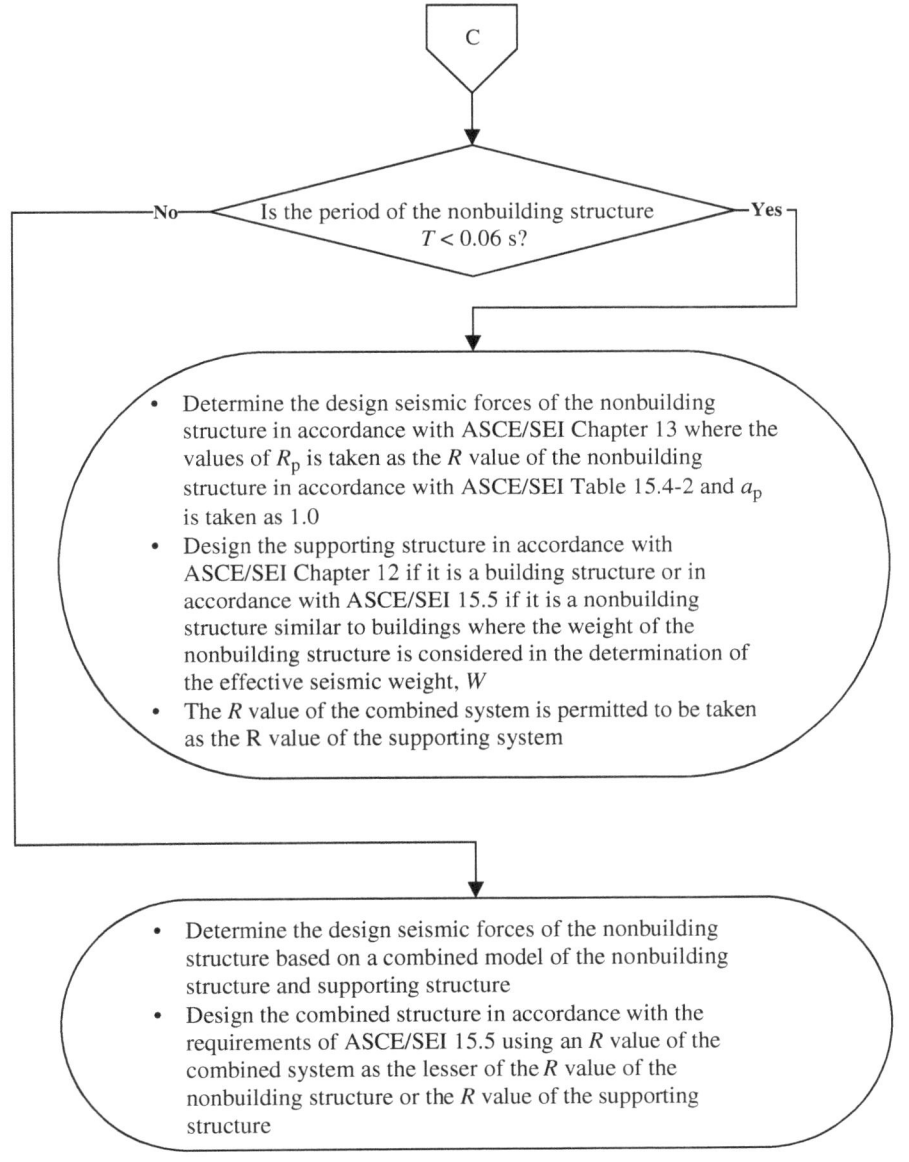

C

Is the period of the nonbuilding structure $T < 0.06$ s?

No   Yes

- Determine the design seismic forces of the nonbuilding structure in accordance with ASCE/SEI Chapter 13 where the values of R_p is taken as the R value of the nonbuilding structure in accordance with ASCE/SEI Table 15.4-2 and a_p is taken as 1.0
- Design the supporting structure in accordance with ASCE/SEI Chapter 12 if it is a building structure or in accordance with ASCE/SEI 15.5 if it is a nonbuilding structure similar to buildings where the weight of the nonbuilding structure is considered in the determination of the effective seismic weight, W
- The R value of the combined system is permitted to be taken as the R value of the supporting system

- Determine the design seismic forces of the nonbuilding structure based on a combined model of the nonbuilding structure and supporting structure
- Design the combined structure in accordance with the requirements of ASCE/SEI 15.5 using an R value of the combined system as the lesser of the R value of the nonbuilding structure or the R value of the supporting structure

FIGURE 5.1   *(Continued)*

5.8   Examples

The following examples illustrate the determination of seismic forces and their effects on nonbuilding building structures based on the requirements given in this chapter.

5.8.1   Example 5.1—Nonbuilding Structure Similar to a Building

Determine the seismic forces on the nonbuilding structure in Fig. 5.2. Design data are given in Table 5.4. The operating weight of each transformer is equal to 30 kips (133.5 kN). The transformers need not remain operational after a design seismic event.

Solution

The flowchart in Fig. 5.1 is used to determine the design seismic forces. It is evident that the nonbuilding structure is similar to a building (see ASCE/SEI Table 15.4-1).

> *Step 1—Determine S_{DS}, S_{D1}, and the SDC*
>
> *Step 1a—Determine S_S and S_1*   ASCE/SEI 11.4.2
>
> In lieu of ASCE/SEI Figures 22-1 through 22-8, S_S and S_1 are determined from Refs. 5, 6, or 7:

$$S_S = 2.092$$

$$S_1 = 0.805$$

> The structure is not permitted to be assigned to SDC A because $S_S > 0.15$ and $S_1 > 0.04$ (ASCE/SEI 11.4.2).
>
> *Step 1b—Determine the site class*   ASCE/SEI 11.4.3
>
> The site class is given as C in Table 5.4.
>
> *Step 1c—Determine S_{MS} and S_{M1}*   ASCE/SEI 11.4.4
>
> Site coefficients F_a and F_v are obtained from ASCE/SEI Tables 11.4-1 and 11.4-2, respectively.

$$\text{For site class C and } S_S = 2.092 > 1.50, F_a = 1.20.$$

$$\text{For site class C and } S_1 = 0.805 > 0.60, F_v = 1.40.$$

> A site-specific ground motion procedure in accordance with ASCE/SEI 11.4.8 need not be performed because the site class is C.

Therefore,

$$S_{MS} = F_a S_S = 1.20 \times 2.092 = 2.510$$

$$S_{M1} = F_v S_1 = 1.40 \times 0.805 = 1.127$$

> *Step 1d—Determine S_{DS} and S_{D1}*   ASCE/SEI 11.4.5

$$S_{DS} = \frac{2}{3} S_{MS} = \frac{2}{3} \times 2.510 = 1.673$$

$$S_{D1} = \frac{2}{3} S_{M1} = \frac{2}{3} \times 1.127 = 0.751$$

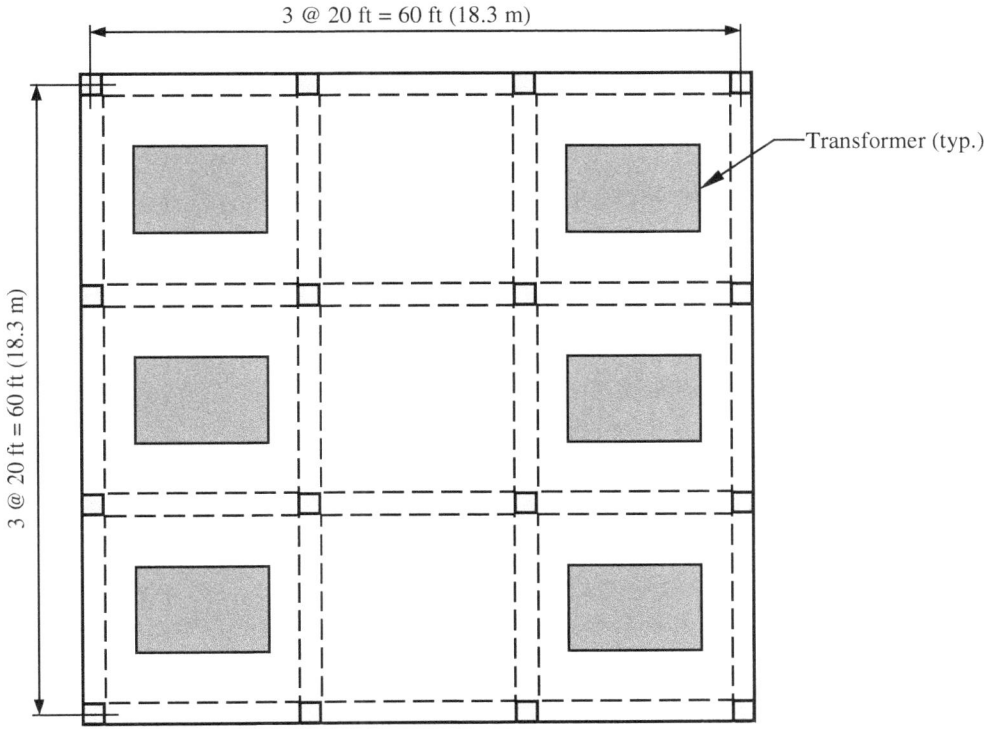

12 in. (305 mm) slab

24 × 24 in. (610 × 610 mm) columns

28 × 24 in. (710 × 610 mm) beams

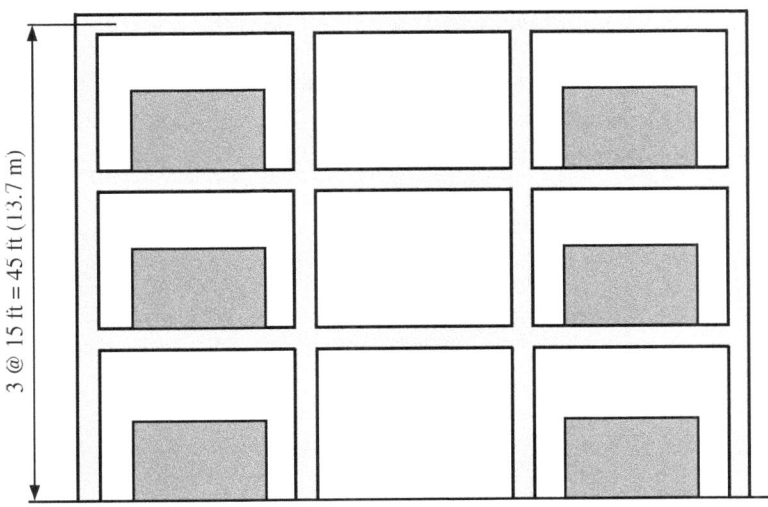

FIGURE 5.2 Plan and elevation of the nonbuilding structure in Example 5.1.

Location	Latitude: 37.870°, Longitude: −122.280°
Soil classification	Site class C
Material	Cast-in-place, reinforced concrete
SFRS	Special reinforced concrete moment frames in both directions
Superimposed dead load on all levels	20 lb/ft² (0.96 kN/m²)

TABLE 5.4   Design Data for the Nonbuilcing Structure in Example 5.1

Step 1e—Determine the SDC   ASCE/SEI 11.6

The risk category is II for this nonbuilding structure.

Because $S_1 = 0.805 > 0.75$, this nonbuilding structure is assigned to SDC E.

Step 2—Determine I_e   ASCE/SEI 15.4.1.1

For this Risk Category II nonbuilding structure, $I_e = 1.00$.

Step 3—Determine the period, T   ASCE/SEI 15.4.4

A three-dimensional analysis of the nonbuilding structure was performed using Ref. 10. From the analysis, the fundamental period, T, is found to be 0.63 s.

Step 4—Determine R, Ω_0, and C_d   ASCE/SEI Table 15.4-1

For special reinforced concrete moments frames:

$$R = 8$$
$$\Omega_0 = 3$$
$$C_d = 5.5$$

Step 4—Determine the design seismic forces   ASCE/SEI 12.6

According to ASCE/SEI Table 12.6-1, the ELF procedure is permitted to be used because the nonbuilding structure does not have any structural irregularities and is less than 160 ft (48.8 m) in height.

Step 4a—Determine the seismic response coefficient, C_s   ASCE/SEI 12.8.1.1

Determine C_s by ASCE/SEI Equation (12.8-2):

$$C_s = \frac{S_{DS}}{\left(\dfrac{R}{I_e}\right)} = \frac{1.673}{\left(\dfrac{8}{1.0}\right)} = 0.209$$

Because $T = 0.63\ s < T_L = 8\ s$, C_s need not exceed that determined by ASCE/SEI Equation (12.8-3):

$$C_s = \frac{S_{D1}}{T\left(\dfrac{R}{I_e}\right)} = \frac{0.751}{0.63 \times \left(\dfrac{8}{1.0}\right)} = 0.149$$

Because $S_1 > 0.6$, minimum C_s is equal to the greater of the values from ASCE/SEI Equations (12.8-5) and (12.8-6) and 0.01:

$$C_s = \text{larger of} \begin{cases} 0.044 S_{DS} I_e = 0.044 \times 1.673 \times 1.0 = 0.074 \\ 0.01 \\ 0.5 S_1 / (R / I_e) = 0.5 \times 0.805 / (8 / 1.0) = 0.050 \end{cases}$$

Therefore, $C_s = 0.149$.

Step 4b—Determine the effective seismic weight, W   ASCE/SEI 15.4.3

The effective seismic weight, W, is the summation of the effective dead loads at each level and includes the weight of the structural members, superimposed dead loads, and the weight of the transformers (details of the weight calculations are not shown here).

The story weights are given in Table 5.5.

Step 4c—Determine the seismic base shear, V   ASCE/SEI Equation (12.8-1)

$$V = C_s W = 0.149 \times 2,896 = 431.5 \text{ kips}$$

In S.I.:

$$V = C_s W = 0.149 \times 12,882 = 1,919.4 \text{ kN}$$

Step 4d—Determine the exponent related to the structure period, k

          ASCE/SEI 12.8.3

Because $0.5 \text{ s} < T = 0.63 \text{ s} < 2.5 \text{ s}$, $k = 0.75 + 0.5T = 1.07$.

Step 4e—Determine the lateral seismic force, F_x, at each level   ASCE/SEI 12.8.3

The force F_x is determined by ASCE/SEI Equations (12.8-11) and (12.8-12). A summary of the lateral seismic forces, F_x, and the story shears, V_x, are given in Table 5.5.

Level	Story Weight, w_x (kips)	Height, h_x (ft)	$w_x h_x^k$	Lateral Force, F_x (kips)	Story Shear, V_x (kips)
R	918	45.0	53,924	212.5	212.5
3	989	30.0	37,646	148.3	360.8
2	989	15.0	17,931	70.7	431.5
Σ	2,896		109,501	431.5	

1 ft = 0.3048 m; 1 kip = 4.4482 kN

TABLE 5.5   Seismic Forces and Seismic Story Shears for the Nonbuilding Structure in Example 5.1

For example, at the second level:

$$C_{vx} = \frac{w_x h_x^k}{\sum w_i h_i} = \frac{989 \times (15.0)^{1.07}}{109,501} = 0.1638$$

$$F_x = C_{vx}V = 0.1638 \times 431.5 = 70.7 \text{ kips}$$

In S.I.:

$$C_{vx} = \frac{4,399.3 \times (4.57)^{1.07}}{136,605} = 0.1637$$

$$F_x = 0.1637 \times 1,919.4 = 314.2 \text{ kN}$$

5.8.2 Example 5.2—Nonbuilding Structure Similar to a Building

Determine the seismic forces on the nonbuilding structure supporting the storage bin in Fig. 5.3. Design data are as follows: $S_1 = 0.60$, $S_{DS} = 1.00$, $S_{D1} = 0.680$, $T_L = 12$ s, and the seismic design category is D.

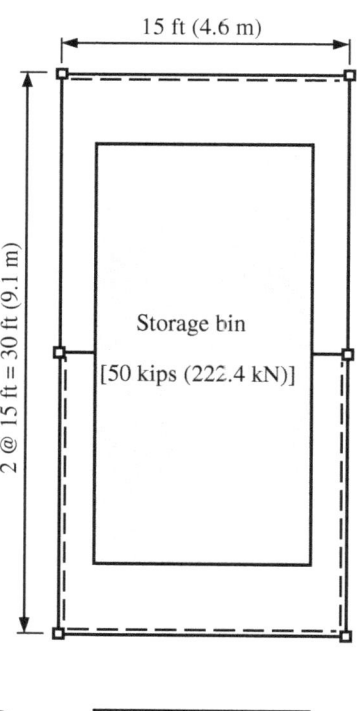

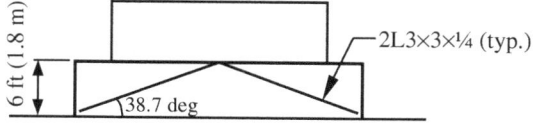

FIGURE 5.3   The nonbuilding structure in Example 5.2.

Solution

The flowchart in Fig. 5.1 is used to determine the design seismic forces. It is evident that the nonbuilding structure supporting the storage bin is similar to a building (see ASCE/SEI Table 15.4-1).

> *Step 1—Determine S_{DS}, S_{D1}, and the SDC*
>
> From the design data, $S_{DS} = 1.00$, $S_{D1} = 0.680$, and the seismic design category is D.
>
> *Step 2—Determine I_e*   ASCE/SEI 15.4.1.1
>
> This nonbuilding structure is assigned to Risk Category II, so $I_e = 1.00$.
>
> *Step 3—Determine the period, T*   ASCE/SEI 15.4.4
>
> In lieu of a more rigorous analysis, T is determined by ASCE/SEI Equation (15.4-6):

$$T = 2\pi \sqrt{\frac{\sum w_i \delta_i^2}{g \sum f_i \delta_i}}$$

The above equation can be reduced to the following for nonbuilding structures with one level:

$$T = 2\pi \sqrt{\frac{w}{gk}}$$

where k is the lateral stiffness of the structure.

The stiffness can be determined by applying a unit horizontal load to the top of the structural frame; this load does not produce any load in the columns. Assuming the elastic shortening of the beams are negligible, only the braces contribute to that lateral stiffness of the structure in the direction of analysis.

From statics, the axial force, u, in each brace due to the unit horizontal load (which is shared between the two sets of braced frames in the direction of analysis) is equal to the following:

$$u = \frac{(1.0/2)}{2 \times \cos 38.7} = 0.32$$

The horizontal deflection, δ, due to the unit horizontal load can be obtained from the following equation (which is from the virtual work method):

$$\delta = \frac{\sum u^2 L}{AE}$$

where L = brace length = $\sqrt{7.5^2 + 6.0^2}$ = 9.6 ft (2.9 m)

A = area of a 2L3 × 3 × ¼ = 2.88 in.² (1,858 mm²)

E = modulus of elasticity = 29,000 kips/in.² (200 kN/mm²)

Therefore, for four braces in the direction of analysis:

$$\delta = \frac{\sum u^2 L}{AE} = \frac{4 \times 0.32^2 \times (9.6 \times 12)}{2.88 \times 29,000} = 0.00057 \text{ in. } (0.01448 \text{ mm})$$

$$k = \frac{1}{\delta} = 1,754 \text{ kips/in. } (307,172 \text{ kN/m})$$

Thus,

$$T = 2\pi \sqrt{\frac{w}{gk}} = 2\pi \sqrt{\frac{50.0}{386.1 \times 1,754}} = 0.05 \text{ s}$$

In S.I.:

$$T = 2\pi \sqrt{\frac{w}{gk}} = 2\pi \sqrt{\frac{222.4}{9.8 \times 307,172}} = 0.05 \text{ s}$$

The weight of the steel is negligible compared to the weight of the storage bin, so it is not included in w.

Because $T = 0.05$ s < 0.06 s, the nonbuilding structure is defined as rigid.

Step 4—Determine the design seismic forces   ASCE/SEI 15.4.2

For rigid nonbuilding structures, the design seismic base shear, V, is determined by ASCE/SEI Equation (15-4.5):

$$V = 0.30 S_{DS} W I_e = 0.30 \times 1.00 \times 50.0 \times 1.0 = 15.0 \text{ kips}$$

In S.I.:

$$V = 0.30 \times 1.00 \times 222.4.0 \times 1.0 = 66.7 \text{ kN}$$

Because this is a one-story structure, this base shear is applied as a concentrated force at the top of the braced frame.

5.8.3 Example 5.3—Nonbuilding Structure Not Similar to a Building

Determine the seismic forces on the 8 in. (203 mm) ground-supported reinforced concrete privacy wall in Fig. 5.4. The wall is anchored into a continuous spread footing. Design data are as follows: $S_1 = 0.34$, $S_{DS} = 0.730$, $S_{D1} = 0.450$, $T_L = 12$ s, and the seismic design category is D.

Solution

The flowchart in Fig. 5.1 is used to determine the design seismic forces. It is evident that this cantilever wall is a nonbuilding structure that is not similar to a building (see ASCE/SEI Table 15.4-2).

Step 1—Determine S_{DS}, S_{D1}, and the SDC

From the design data, $S_{DS} = 0.730$, $S_{D1} = 0.450$, and the seismic design category is D.

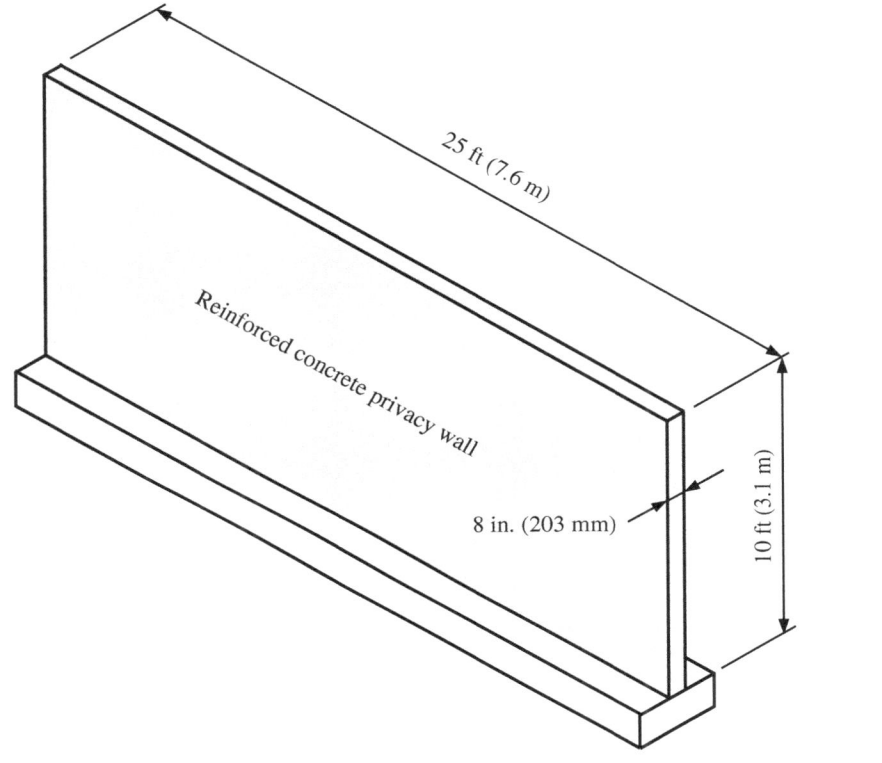

Figure 5.4   The reinforced concrete privacy wall in Example 5.3.

Step 2—Determine I_e   ASCE/SEI 15.4.1.1

This wall is adjacent to a residential complex, so the risk category for the wall is II. Thus, $I_e = 1.00$.

Step 3—Determine the period, T   ASCE/SEI 15.4.4

From a dynamic analysis of the cantilever wall, $T = 0.20$ s > 0.06 s.

Step 4—Determine R, Ω_0, and C_d   ASCE/SEI Table 15.4-2

For ground-supported cantilever walls:

$$R = 1.25$$
$$\Omega_0 = 2$$
$$C_d = 2.5$$

Step 4—Determine the design seismic forces   ASCE/SEI 15.1.3

According to ASCE/SEI 15.1.3, the ELF procedure in ASCE/SEI 12.8 is permitted to be used to determine the design seismic forces for nonbuilding structures that are not similar to buildings subject to the requirements in ASCE/SEI 15.4.

Step 4a—Determine the seismic response coefficient, C_s   ASCE/SEI 12.8.1.1

Determine C_s by ASCE/SEI Equation (12.8-2):

$$C_s = \frac{S_{DS}}{\left(\dfrac{R}{I_e}\right)} = \frac{0.730}{\left(\dfrac{1.25}{1.0}\right)} = 0.584$$

Because $T = 0.20$ s $< T_L = 12$ s, C_s need not exceed that determined by ASCE/SEI Equation (12.8-3):

$$C_s = \frac{S_{D1}}{T\left(\dfrac{R}{I_e}\right)} = \frac{0.450}{0.20 \times \left(\dfrac{1.25}{1.0}\right)} = 1.80$$

According to ASCE/SEI 15.4.1(2), minimum C_s is equal to the following:

$$C_s = \text{larger of} \begin{cases} 0.044 S_{DS} I_e = 0.044 \times 0.730 \times 1.0 = 0.032 \\ 0.030 \end{cases}$$

Therefore, $C_s = 0.584$.

Step 4b—Determine the effective seismic weight, W   ASCE/SEI 15.4.3

The effective seismic weight, W, is the weight of the wall:

$$W = \frac{8.0}{12} \times 150 \times 10.0 \times 25.0 / 1{,}000 = 25.0 \text{ kips}$$

In S.I.:

$$W = 203 \times 23.6 \times 3.1 \times 7.6 / 1{,}000 = 112.9 \text{ kN}$$

Step 4c—Determine the seismic base shear, V   ASCE/SEI Equation (12.8-1)

$$V = C_s W = 0.584 \times 25.0 = 14.6 \text{ kips}$$

In S.I.:

$$V = C_s W = 0.584 \times 112.9 = 65.9 \text{ kN}$$

This force is applied at the center of mass of the wall.

CHAPTER 6

References

1. International Code Council. 2017. *2018 International Building Code*, Washington, DC.
2. Structural Engineering Institute of the American Society of Civil Engineers (ASCE). 2017. *Minimum Design Loads and Associated Criteria for Buildings and Other Structures*, ASCE/SEI 7-16 including Supplement 1, Reston, VA.
3. Building Seismic Safety Council (BSSC). 2015. *NEHRP Recommended Seismic Provisions for New Buildings and Other Structures*, FEMA P-1050-1, Washington, DC.
4. International Code Council. 2017. *2018 International Residential Code*, Washington, DC.
5. American Society of Civil Engineers (ASCE). 2020. ASCE 7 Hazard Tool. https://asce7hazardtool.online/.
6. Applied Technology Council (ATC). 2020. ATC Hazards by Location. https://hazards.atcouncil.org.
7. Structural Engineers Association of California (SEAOC) and Office of Statewide Health Planning and Development (OSHPD). 2020. Seismic Design Map Tool. https://seismicmaps.org/.
8. Building Seismic Safety Council (BSSC). 2016. *2015 NEHRP Recommended Seismic Provisions: Design Examples*, FEMA P-1051, Washington, DC.
9. Chopra, A.K. 2017. *Dynamics of Structures*, Prentice Hall, Upper Saddle River, NJ.
10. Computers and Structures, Inc. (CSI). 2020. ETABS—Integrated Analysis, Design and Drafting of Building Systems, Version 16.2.1, Walnut Creek, CA.
11. American Concrete Institute (ACI). 2019. *Building Code Requirements for Structural Concrete and Commentary*, ACI 318-19, Farmington Hills, MI.
12. The Masonry Society (TMS). 2016. *Building Code Requirements and Specification for Masonry Structures*, TMS 402/602-16, Longmont, CO.

Ing... ...roup UK Ltd.
Milton ... UK
UKHW03 ...21190323
418794UK00007B/213